Adobe® Acrobat® and PDF for Architecture, Engineering, and Construction

W0036347

Tom Carson and Donna L. Baker

Adobe® Acrobat® and PDF for Architecture, Engineering, and Construction

With 255 Figures

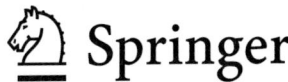 Springer

Tom Carson, P.E. ACE
New Economy Institute
1105 E. 10th Street
Chattanooga
TN 37403
USA

Donna L. Baker, ACE
24 Davidson Place
St. Andrews
Manitoba
Canada

British Library Cataloguing in Publication Data
Carson, Tom
    Adobe Acrobat and PDF for architecture, engineering, and
    construction
    1. Adobe Acrobat (Computer file) 2. PDF (Computer file
    format) 3. Architecture - Data processing 4. Engineering -
    Data processing 5. Building - Data rocessing
    I. Title II. Baker, Donna L., 1955-
    005.7'2
ISBN-10: 1846280206

Library of Congress Control Number: 2005932864

ISBN-10: 1-84628-020-6   e-ISBN: 1-84628-138-5   Printed on acid-free paper
ISBN-13: 978-1-84628-020-7

© Springer-Verlag London Limited 2006

Adobe, Acrobat, Acrobat Reader, Adobe Reader, Distiller, Illustrator, InDesign, LiveCycle, LiveCycle Designer, Photoshop, Photoshop Album, PostScript and Reader are either registered trademarks or trademarks of Adobe Systems Incorporated in the United States and/or other countries.

Autodesk, AutoCAD and Autodesk Inventor are either registered trademarks or trademarks of Autodesk, Inc., in the USA and/or other countries.

"GeoPDF" © Layton Graphics, 2005, All rights reserved.

MasterFormat is the trademark of the Construction Specifications Institute (in the USA) and of Construction Specifications Canada (in Canada).

All other brand names, product names, or trademarks belong to their respective holders.

THIS PRODUCT IS NOT ENDORSED OR SPONSORED BY ADOBE SYSTEMS INCORPORATED, PUBLISHER OF ADOBE ACROBAT.

Apart from any fair dealing for the purposes of research or private study, or criticism or review, as permitted under the Copyright, Designs and Patents Act 1988, this publication may only be reproduced, stored or transmitted, in any form or by any means, with the prior permission in writing of the publishers, or in the case of reprographic reproduction in accordance with the terms of licences issued by the Copyright Licensing Agency. Enquiries concerning reproduction outside those terms should be sent to the publishers.

The use of registered names, trademarks, etc. in this publication does not imply, even in the absence of a specific statement, that such names are exempt from the relevant laws and regulations and therefore free for general use.

The publisher makes no representation, express or implied, with regard to the accuracy of the information contained in this book and cannot accept any legal responsibility or liability for any errors or omissions that may be made.

Printed in Germany

9 8 7 6 5 4 3 2 1

Springer Science+Business Media
springeronline.com

This book is dedicated to the spirit of engineering embodied in the Engineering (Tennessee River) Valley – from harnessing the atom, to developing fertilizers to feed the world, to building rockets to the moon – the spirit has been a constant motivator in my career.

Foreword

This book is for those who have a seemingly endless inflow of data on paper, drawings, spreadsheets, emails, and electronic documents in different software formats. In short: all of us! This book is written for those who desire success in a new economy that is driven by knowledge engineering and management and fueled by exponentially advancing technology.

Our management strategies to cope have brought us further complications. The current trend toward outsourcing means more incompatible data formats, as well as more paper that must somehow be integrated into meaningful information.

Tom and Donna have written a reference to simplify our increasingly complex technical lives in a style of reading that immediately provides a roadmap to integrate all sources of data into information and then knowledge.

This is not another book simply explaining how to use the new version of Adobe Acrobat. It does that nicely, but the real value is the self-learning information system outlined for your success.

Gregory A. Sedrick, Ph.D., P.E.
The New Economy Institute
February 2005

Preface

Tom Carson's professional career was as a civil/environmental engineer in hazardous waste site remediation and hazardous materials management. He was instrumental in the early days of the Certified Hazardous Materials Managers Program, and co-edited the Handbook on Hazardous Materials Management. Engineering is a very paper-intensive industry; engineering with hazardous materials and waste is even more so.

After 20 years in the field, Tom went in search of ways to make engineering more efficient, with the goal of conquering the paper dragon. Adobe Acrobat 3 was the version Tom first discovered, and he quickly realized the program's potential for solving the engineering paper problem. Self-teaching, experimenting, and his engineering background have helped Tom create and explore new uses for Acrobat.

Following a stint in the Canadian army, Donna Baker became a nurse, working in the far north, often being flown in to help deliver a problem baby or work with a seriously sick patient. Once Donna had a family, she completed business and graphic arts degrees, while teaching in a business college. One of her later business tasks was creating technical materials for a software company, which honed her writing skills.

One of Donna's first computer books was *Adobe Acrobat 5: The Professional User's Guide*. Tom bought all the books on Acrobat 5 and judged her book the best. After exchanging many emails and phone calls, Tom begged his way to become co-author of *Adobe Acrobat 6: The Professional User's Guide*.

Donna has written several other Acrobat books, in addition to numerous other titles. With her *Adobe Acrobat 6 Tips and Tricks* book winning an International Award of Excellence from the Society of Technical Communications in 2005, she has become the "Queen of PDF Books."

This book combines Tom's experience with engineering and Acrobat, and Donna's Acrobat expertise and writing ability. Applying the information in this book can significantly improve engineering efficiency – and save a forest or two.

Tom Carson P.E. ACE   Donna L. Baker A.C.E.

Acknowledgements

We would like to thank the City of Chattanooga Tennessee Parks, Recreation, Arts & Culture Department and March Adams and Associates Consulting Engineers of Chattanooga for use of the files for the DuPont Soccer Complex.

Tom Carson

I would like to thank the New Economy Institute and the Southeast Local Development Corporation for allowing me to develop my obsession with PDF into interesting projects that are helping the area.

I would also like to thank my wife Judy for doing my chores so I could write this book. I know she is glad that it is over before mowing season.

And thank you to the "Queen of PDF Books." I would never have tackled this without her.

Donna L. Baker

I would like to thank my pal Tom Carson for inviting me to work on this project. It has been my pleasure and a great learning experience. Well, lots of work too, but extremely interesting. Thank you to Oliver Jackson and Anthony Doyle at Springer for their assistance and patience.

Thanks to my husband Terry, and to my girl Erin for keeping me in touch with a nondigital reality. Finally, my thanks as always to Tom Waits for singing to my spirit.

Table of Contents

Contributors

We are grateful to those who have contributed of their time and expertise to enhance the quality and expanse of the information in our book.

Michael Bufkin

Michael Bufkin has been involved in CAD and Image Processing for engineering since 1979. In 1983 he founded CADDShare Corporation, a CAD software developer, which was acquired by Layton Graphics in 1992. Since then he has served as CTO and has led Layton Graphics in its transition from a microfilming service bureau to it current position as a leading developer of engineering and mapping applications for Adobe Systems users. Mr. Bufkin holds an MS in Engineering from The University of Texas, where he was a National Science Foundation Fellow. He is the author of numerous papers on CAD and image processing.

Michael contributed the chapter on geoPDF and georeferencing.

Tim Huff

Tim Huff received a BSME University of Texas. Before going into the software realm he worked for General Dynamics on the flight dynamics of the F22 and F117 then moved to Foster Wheeler where he designed process plants and towers. Mr. Huff then moved to AutoDesk, where he spent time in Field Sales and as NA AE manager, then took the position of Product Manager for the Inventor Product. After 8.5 years, Mr. Huff left and went to Solid Works to drive their World Wide field marketing team and product manager for core Solid Works. Mr. Huff is now at Adobe as the business development manager for Acrobat in AEC.

Tim contributed information on creating AutoCAD PDF files.

Glenn McLain

Glenn has been involved in the aerospace and aviation industries for many years, with an educational background in civil engineering, business management, and aviation. He has many years of aviation experience with transport category aircraft, and at one time owned an aviation/computer consulting company. Glenn has managed a variety of technical publications, and been instrumental in the development of various information repositories and manuals in aerospace and related industries.

Glenn served as technical reader for this project.

Lyn Price

Lyn Price is the Assistant Director of The New Economy Institute in Chattanooga, TN. She has worked in economic and business development for the past 5 years. She has a BS in Mathematics from the University Tennessee in Knoxville, TN. She has been working with Acrobat for the past 2 years – since she started working with Tom. Lyn currently lives in Chattanooga with her husband, Sean, and two children, Allison and Brandon.

Lyn contributed the chapter on working with Acrobat forms.

Jo Terri Wright

Jo Terri Wright is a publishing expert in the Publishing Products group at Bentley Systems, Inc. in Huntsville, Alabama. She is responsible for technical marketing of the publishing solutions in the MicroStation and ProjectWise product lines. Ms. Wright has worked at Bentley Systems since 2000, when the InterPlot product group was acquired from Intergraph Corporation. At Intergraph she has held various roles, including on-site support of the Intergraph systems for NASA at the Marshall Space Flight Center; support, certification and training for the scanning division; and pre-sales and marketing for the workstation and plotting divisions. Ms. Wright is a member and Co-Chairman of the PDF-E Committee formed to establish an ISO standard for PDF documents for engineering.

Jo Terri contributed the information on working with MicroStation, as well as using 3D PDF files.

1

Introduction

Once every few generations something new appears that makes quantum leaps in Architecture, Engineering and Construction (AEC). Paper was a major leap – it certainly beat carrying around clay tablets! The slide rule was another quantum leap. As late as 1960, computers were people that computed with a slide rule; with these computers we were able to put a man on the moon.

Machines that did the computing resulted in many engineering breakthroughs, but for all the major power, paper was still the only common format for communication. Adobe© Acrobat© and Portable Document Format (PDF) is the next jump forward in AEC, providing the first common format since paper.

We will never be completely free of paper, but the common format allows us to do amazing things, including:

- Create electronic bid sets complete with drawings, specifications and fillable forms.
- Submit bids electronically on a PDF form that contains all the math, but no math errors.
- Submit and manage all submittals in the PDF format.
- Electronically redline drawings and specifications on-line.
- Search thousands of drawings and manuals at the same time.
- Create geo-referenced PDF maps with embedded data at geo-spacial points.

Does this list read like a software salesman promising the moon? It definitely is not!

The Development of PDF

The PDF format was invented by Dr. John Warnock of Adobe Systems. Adobe had invented the PostScript© printer driver, which helped create the mountain of paper under which we are smothering. About 12 years ago, John modified the PostScript printer driver to print to an electronic space instead of to paper, which allowed for file sharing between the Macintosh and PC platforms, as well as the transfer of files from Adobe products to commercial printers.

Dr. Warnock's decision to freely distribute the Acrobat Reader© was unheard of at the time, as was his decision to make the specification available, freely encouraging others to develop plug-ins to extend the capabilities of Acrobat©. Adobe spent a fortune on developing PDF, so how could you make money giving away a program?

Further, who would encourage competitors to build plug-ins, and encourage competitors to develop with Adobe intellectual capital? John proved to be crazy as a fox. Over 750 million copies of the Acrobat Reader and Adobe Reader© have been downloaded. The reader's name was changed to "Adobe" Reader at the release of Acrobat 6 to avoid confusion between the software and reader products.

Some of the brightest programmers in the world have added to the Adobe product by developing added functionality, making Acrobat and its Reader much greater than the sum of its parts.

Build It and They Will Come – PDF Standards

A major problem with the computer world has been the lack of standardization. PDF is becoming the standard for multiple industries. The International Organization for Standardization (ISO) has created, or is currently developing, several standards for PDF.

PDF/X (PDF Graphics eXchange) is a series of standards for the commercial printing industry.[1] Files adhering to these standards are guaranteed to produce printed files exactly as the creator intended. Although this standard is of limited value to the AEC world, it started the development of standards in different arenas of PDF use.

PDF/A Standard

Another PDF standard, PDF/A (PDF Archival) is currently in development. Several US agencies, including the US Federal Court System and the US National Archives and Records Administration (NARA), have been instrumental in motivating the development of the PDF/A standard.

An ISO committee was formed to set standards for PDF files that would be usable 50 years in the future. The standard will set minimum resolution for raster images, determine features that will not be acceptable, such as links to URLs that will probably be long gone, or embedding of multimedia formats that in all likelihood will be obsolete. The standard requires that only non-proprietary formats be used. The PDF/A standard will be released by the time this book is printed.[2]

PDF/E Standard

The AEC world has been extremely fragmented. While there are several very good computer-aided design (CAD) and geographical information system (GIS) programs, there is no single program that is good for all branches of AEC. As a result, we are faced with numerous proprietary formats, each requiring a proprietary viewer. It is true that some of these formats may have a few features that are superior to current PDF features, but the ability to bring everything together in one format greatly exceeds the benefits of these select few features.

Several of the greatest engineering groups in the world are working together to produce an ISO Standard PDF/E (PDF Engineering) which will adapt the current standard to meet AEC needs. The preliminary discussions and development committees are formed under the auspices of the Association for Information and Image Management (AIIM).[3]

---

[1] See ISO 15929:2002 [1]
[2] See ISO/DIS 19005-1 [2]
[3] PDF-Engineering Committee, AIIM [3]

The standard will likely open the PDF standard for 1:1 scale drawings. Three-dimensional (3D) PDF documents will be possible under this standard, as will animated drawings that interactively show the consequence of an action, such as removal of a critical part. The goal of the standard is to improve document exchange, collaboration, and print accuracy within engineering workflows. PDF/E will allow the AEC community to design in the most appropriate formats, but manage the complete project in one electronic PDF format.

The Current State of PDF

The intricate design/build/operate phases of the AEC lifecycle encompass a mix of related information, including 3D models, two-dimensional (2D) drawings, and written specifications. Document sets serve as the currency of AEC projects and become the books of record for large, as-built assets. AEC is a $3.4 trillion industry where documentation is the basis for product and services delivery.[4]

Architects, engineers, and contractors require a safe, small, smart format to share and archive these sets globally.

Using PDF

Engineering workflows are complicated, often entailing thousands of drawings, specification sheets, notations, and other materials. Maintaining and updating even a small project can be an onerous task. Using Acrobat and a logical workflow, a project can be managed more effectively.

Imagine being able to search all the drawings of a spacecraft for a defective bolt: Currently we can use drawings from several CAD programs converted to PDF. In some circumstances, we can search thousands of CAD drawings in seconds. The search can include CAD files originally drawn with fonts, provided those fonts are true scaleable fonts (and therefore searchable), or rendered with Adobe Acrobat 7.

Several utilities are using geo-referenced PDF files with embedded data working with MAP2PDF, a product developed by Layton Graphics.[5] Working with the free Adobe Reader and a plug-in or two, field crews can access all the system's information. For example, a click on a power pole will reveal embedded data such as the type, height, associated utilities, and other equipment. A map displaying the exact location can be found in seconds. A field supervisor using Adobe Acrobat can field annotate changes for the engineering staff to update the GIS and engineering databases. A server automatically publishes up-to-date PDF maps.

Industry and Government Adoption

PDF is the standard format for submission and distribution in government and regulatory agencies worldwide. To date, there are sales of over 14 million seats of the full Acrobat product. Over three-quarters of a billion copies of Acrobat Reader and Adobe Reader have been licensed, and the Reader is bundled with 60% of all PDAs, as well as distributed by the top ten PC manufacturers.

---

[4] Industry value quoted in article derived from McGraw-Hill Construction sources. [4]
[5] MAP2PDF converts GIS data formats to intelligent geo-referenced non-proprietary PDF files. [5]

Welcome to our Book

The AEC fields are under constant pressure to work quicker, better and cheaper. Through the processes identified in our book, Acrobat will help you achieve this goal by realizing time and cost savings.Areas in which you can apply the knowledge in this book include:

- Assembling and converting source materials for the bid package. Paper copies are expensive to print and distribute.Working with bid sheets. Well-designed PDF forms prevent errors and save time in bid tabulation.
- Making the submission. Paper submittals are expensive and time consuming. Acrobat electronically automates and manages the process, saving days!
- Project closeout. Instead of rolls of redlines and boxes of manuals, the project becomes an electronic Owner's Manual that lasts throughout the finished product's lifetime.

Our book is aimed at a professional audience, both practitioners and students, in a variety of engineering, architecture, and construction sectors. PDF has been a *defacto* standard in document sharing and distribution for some time.

Table 1.1 identifies a number of common disciplines and how relevant PDF is to that discipline. We have also outlined specific features in PDF and this book that are of particular importance to the different disciplines.

Learning for the Future

This book will prepare you for the next AEC quantum leap. You will have to learn the fundamentals of PDF and Acrobat to do the job, just as you did to learn basic engineering. The book is designed to teach Acrobat within the framework of a design-build environment as you follow the processes in an AEC project.

You will prepare a bidset using the Acrobat PDFMakers©, macros installed into a number of programs when Acrobat is installed, as well as the Adobe PDF Printer©, a printer driver that generates PDF files from print commands.

Acrobat 7 Professional includes two different ways to produce forms. You can work with the Forms tools within Acrobat, or use Adobe Designer, a dedicated forms program accessible directly from within Acrobat.

In the book, you will learn how to make a simple survey form in Adobe Designer. Using the knowledge, you can design your own project forms, such as an electronic bid submittal form that can even be linked to a database to identify irregularities in the bidding process.

Adobe has taken the paper model for submittal review and tracking and automated it electronically. Acrobat coordinates the review process using the Tracker and a great set of commenting tools. A major new feature for users of Acrobat 7 Professional is the ability to enable users of Adobe Reader to use Drawing Markups and Commenting tools, effectively expanding access to PDF tracking of documents to millions of additional users. This new feature ties in external users that do not have access to the full program. Adobe is providing a big feature for free.

Table 1.1 Importance of PDF to AEC disciplines

Discipline	Relative Importance	PDF Features
Electrical Engineering	85%	Review tracking; project management
Civil Engineering	90%	Geo-referenced drawings; project management
Mechanical Engineering	85%	All CAD is common format; configuration management
Construction Managers	95%	Reduced review and approval time; better document control
Architects	95%	Easier to work with clients, since all files use a common format; easier to work with permitting agencies
Knowledge Managers	100%	All information in one format and completely searchable
Researchers	75%	Currently being used for knowledge and research management

How this Book is Presented

You will learn to use Acrobat and PDF working through the book. The book generally follows the design build process.

- **Introduction.** The introductory chapters describe how engineering processes are defined for the purposes of the book's discussions.
- **Converting Source Materials for the Bid Package.** You see how to convert files of different types of PDF using a variety of methods, as well as how to work with images and other file formats.
- **Assembling Source Materials for the Bid Package.** Learn how to assemble a document, and add navigation and security features.
- **Working with Bid Sheets.** Discover how to use Acrobat forms tools, as well as how to work with Adobe Designer.
- **Making the Submission.** This part is all about communication. You learn how to initiate, track, and conduct a review. You also see how to work with the many viewing and commenting tools at your disposal.
- **Project Closeout.** In this part, learn how to prepare document collections, build an index, and repurpose content in a PDF document.
- **Customizations and Advanced Features.** The final part of the book includes information on a variety of other program functions that you can use to make your workday more productive and enhance your projects. You will learn how to use multimedia, embedded data, and geo-referenced data.

Dupont Soccer Complex

To put your learning into practice, we include a cumulative project threaded through most of the chapters in this book. The project is composed of an abbreviated series of files from real projects that have been modified to teach Acrobat and the elements of AEC electronic project management.

We would like to thank the City of Chattanooga Tennessee Parks, Recreation, Arts & Culture Department and March Adams and Associates Consulting Engineers of Chattanooga for use of the files for the DuPont Soccer Complex.

Note: Many elements of the book's project have been modified due to copyright and liability issues, and should not be used in other projects. They are only being used to teach the processes within the context of this book.

References

[1]   ISO 15929:2002 (2002), Graphic technology-Prepress digital data exchange-Guidelines and principles for the development of PDF/X standards, International Standards Organization, Geneva, Switzerland.

[2]   ISO/DIS 19005-1 (2004), Document management – Electronic document file format for long-term preservation – Part 1: Use of PDF 1.4 (PDF/A-1), PDF-Archival International Standards Organization, Geneva, Switzerland.

[3]   AIIM, AIIM, The ECM Association - PDF-Engineering, n.d. http://www.aiim.org/standards.asp? ID=27860.

[4]   PlanetPDF, *Bentley and Adobe Team to Drive PDF as a Standard for AEC, 27 Oct. 2003.* http://www.planetpdf.com/mainpage.asp?webpageid=3178.

[5]   Layton Graphics, *MAP2PDF Product Information. 29 Dec. 2003.* http://www.layton-graphics.com/map2pdf.html.

2

The AEC Workflow

You won't find anything new in the first part of this chapter if you are a gray-haired AEC professional, as you have gained experience working with a paper-based workflow the hard way. If you are a student, carefully read about paper-based workflows in the first part of this chapter.

Regardless of your level of expertise, the second part of the chapter describes a new way of working – without paper. Acrobat offers an electronic workflow that effectively replaces the transport and processing of multiple iterations of files and documents.

In this Chapter

The AEC workflow is based on a series of coordinated and interconnected events. In this chapter you will learn about the components of an AEC workflow, first by examining a paper-based workflow, and then an equivalent digital workflow

You will learn that:

- All projects start with a definition of scope, a document on which the design is formulated.
- The design process defines the structure of a project, and the contributions made by various specialties and disciplines.
- The bid is developed using a methodical, systematic formatting system, and submitted.
- The bid opening announces the winner of the bid, as well as qualifying the winner.
- The submittal process defined in the contract requires that documents identifying materials used in the project be submitted to the owner for review.
- According to the contract's requirements, payments are generated based on specified milestones; change orders are processed on an *ad hoc* basis and subject to review and approval.
- Project wrap-up includes provision of sets of redlined drawings, as well as manual sets for all equipment.

A Paper-based AEC Workflow

For purposes of discussion, we are presenting two different workflow diagrams in this chapter, both developed by Adobe Systems.

Adobe has developed a highly simplified, but very true diagram of a paper-based AEC workflow (Figure 2.1). Everywhere you see an airplane, mail truck, or fax machine in the figure, you can also add a dollar ($) sign. The cost is not only in the actual costs of producing the printed pages and moving the paper – much of the cost is in the time invested in the distribution of the paper.

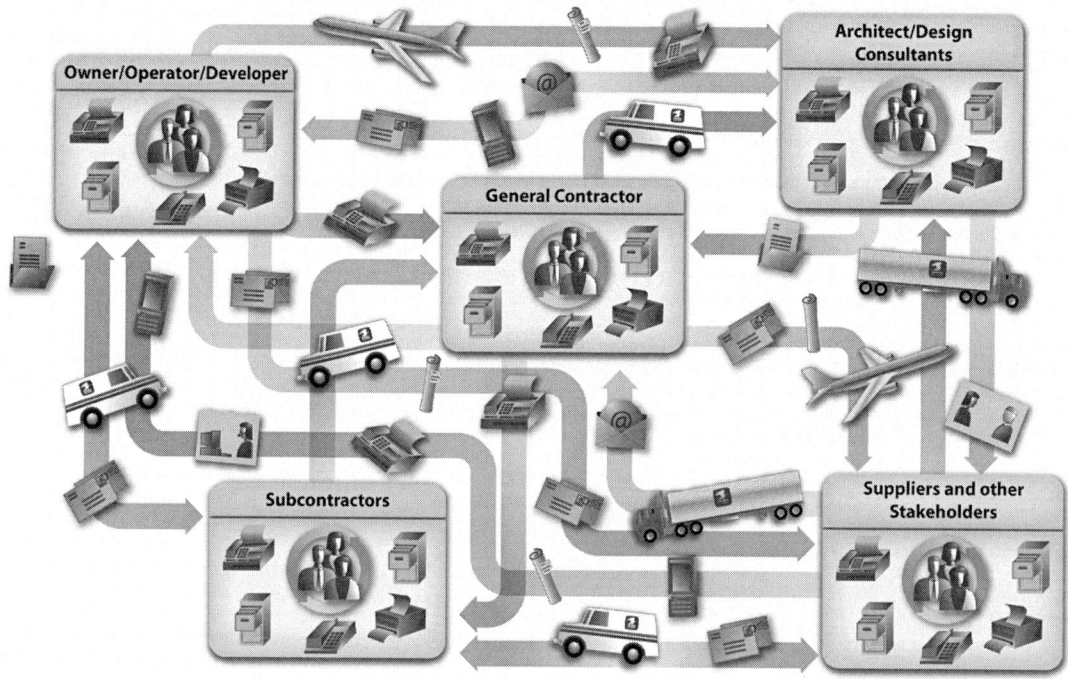

© 2004 Adobe Systems Incorporated. All rights reserved. Adobe and Acrobat is/are either [a] registered trademark[s] of Adobe Systems Incorporated in the United States and/or other countries.

Figure 2.1 A complex, paper-based workflow

Project Scoping

Let us take a look at the workflow in detail. A project starts with an idea from the Owner/Operator/ Developer. Either the project initiator prepares a scoping document identifying their needs, or the task is assigned to the Architect/Design Consultant (A/DC). The needs and constraints of the owner must be detailed in a format that the A/DC can use for generating design parameters. Producing a working document may require multiple iterations, and a lot of paper is moved during project scoping.

Design Process

The design process begins once the project scope is approved. The workflow diagram shows a single box for the A/DC, although the vast majority of projects are not done via a single entity.

Some very large architecture and engineering firms have all disciplines in house, but are the exception rather than the rule. Even the large A&E firms grow by acquisition and may have the disciplines needed at different locations, using incompatible design programs. Coordinating design efforts among disciplines requires considerable paper shuffling and distribution using various mediums. There are a lot of invisible $ signs in the design process.

The Bid Process

The specifications on a project are designed using MasterFormat^{TM} Numbers and Titles, a system developed jointly by the Construction Specifications Institute and Construction Specifications Canada that has been in use since the early 1960s. The basic divisions, groups, and subgroups are listed in Table 2.1.

MasterFormat™ is the specifications-writing standard for most North American commercial building design and construction projects. The standard lists titles and division numbers to organize data about products, construction requirements, and activities [1]. The system is used to standardize filing and retrieval schemes, and is appropriate for many types of communication, such as project manuals, cost and technical data, reference notes on drawings, and product information.

The MasterFormat^{TM} standard also contains up to 999 subdivisions, as well as cross-references, which are important when an item could logically be placed in multiple locations. *See* and *See Also* references direct the user to other divisions or subdivisions.

Plans and specifications are prepared and, since paper has been the only common format, they are printed. The cost to produce a set of plans and specifications can easily reach or exceed $100. In the past, the cost of producing the plans and specifications was covered by the owner, but no-cost plans often resulted in too many people requesting plans that did not submit bids. Some owners started requiring a deposit on the plans and specifications, while others pre-qualified the bidders. Pre-qualification of bidders adds time to the process, and commands higher fees to the designers to qualify the bidders. Many owners opt for requesting a set fee for a copy of the plans and specifications to offset the printing costs.

Plans Based on Discipline

Plans are often arranged around discipline and may change based on the discipline in charge. Let us look at an example using site work. If an engineer is in charge, the pages may be numbered starting at C-01 (for Civil). If an architect is in charge, the pages may be numbered starting at AS-01 (for Architectural Site).

For the most part, the page number prefixes follow this pattern: "A" for Architecture, "C" for Civil, "E" for Electrical, "M" for Mechanical, "S" for Structural, and "P" for Plumbing. Other work may be prefaced with "MP" for Mechanical Process and "NC" for Controls and Instrumentation.

Developing the Bid

The general contractor rarely has all the specialties in house, and sends plans and specifications to subcontractors and vendors for price quotations. The contractor often spends the time and money to make a complete subset of the plans and specifications for each of the mechanical, plumbing, electrical, civil, or specialty contractors providing a bid.

Creating subsets of plans and specifications can save money on reproduction, but may cost the contractor if a key piece of information is hidden in part of the specifications not furnished to the subcontractor. A contractor may need a dozen copies of a project's plans and specifications to distribute to the various subcontractors.

Table 2.1 MasterFormat™ specification divisions

Group	Division	Number
Procuring and Contracting Requirements	Procurement and Contracting requirements	00
Specifications Group	General Requirements	01

Facility Construction Subgroup

Division	Number	Division	Number
Existing Conditions	02	Finishes	09
Concrete	03	Specialties	10
Masonry	04	Equipment	11
Metals	05	Furnishings	12
Wood, Plastics, and Composites	06	Special Construction	13
Thermal and Moisture Protection	07	Conveying Equipment	14
Openings	08		

Facility Services Subgroup

Division	Number	Division	Number
Fire Suppression	21	Electrical	26
Plumbing	22	Communications	27
HVAC	23	Electronic Safety and Security	28
Integrated Automation	25		

Site and Infrastructure Subgroup

Division	Number	Division	Number
Earthwork	31	Transportation	34
Exterior Improvements	32	Waterway and Marine Construction	35
Utilities	33		

Process Equipment Subgroup

Division	Number	Division	Number
Process Integration	40	Pollution Control Equipment	44
Material Processing and Handling Equipment	41	Industry-specific Manufacturing Equipment	45
Process Heating, Cooling, and Drying Equipment	42	Electrical Power Generation	48
Process Gas and Liquid Handling, Purification, and Storage Equipment	43		

The Bid Opening

The bid is developed and delivered or faxed to the bid opening. The apparent winner is often announced at the opening. However, the winning bid is not official until the engineer reviews the bids and determines that the apparent winner is in the owner's best interests.

The determination process usually involves manually developing a spreadsheet, adding data from competing bids, and then comparing the bids. Often, a contractor can detect a bad assumption on the engineer's part and load the bid to make a certain line item and the total bid low. The contractor is then in a position for a change order and renegotiation of the line item. The engineer can catch the assumption during the determination process, and decide the bid is not in the best interests of the owner.

The Submittal Process

Once the project is awarded, the contractor is responsible for submittals required by the contract. The submittals assure the owner's representative that the material used is the same or equal to the specified materials, and that choices of materials and methods will not cause problems with other parts of the project.

The submittal process is highly resource intensive. Up to six copies of each submittal are required, including a form. The submittal form tells the engineer where the submittals originate in the project's specification.

Once received, the copies are logged and distributed to the reviewers by the document manager, who is usually a junior engineer. The reviewers mark up the documents and return comments to the document manager. The document manager then forwards them to the engineer-in-charge, who reviews the comments and approves, conditionally approves, or rejects the submittal. The submittal is returned to the document manager again, who sends the results back to the contractor.

The Payment and Change Order Processes

Each project's documentation includes forms for payment and for change orders. Paper forms are commonly used, and of course it is simple to make calculation errors on the forms.

The contractor fills out the forms after calculating the math, and the project engineer double-checks the figures. According to the contract's requirements, payments are generated based on specified milestones; change orders are processed on an *ad hoc* basis and subject to review and approval.

Project Wrap-up

One of the last things that is done on a project is to generate a set of redlined drawings and manuals for all the equipment. *Redlined* drawings are project drawings on which the contractor identifies the changes with a red pencil. Often, a contract requires two sets of redlines for both the owner and the engineer; the contractor is remiss in not making a third set of redlined drawings for himself. Until recently, large-scale color scanners were prohibitively expensive and redlining on all sets was added by hand.

In addition to redlined drawings, most contracts usually require two sets of equipment manuals. Some contracts require that the manuals be organized in binders, while other contracts do not require that level of organization.

Future Considerations

Suppose it is one year later, and the contractor is called back to the site to work on the equipment. He may or may not be able to find the drawings or manuals, and even if he does, it is not necessarily the case that the drawings will be updated based on the contractor's work, thus establishing a scenario that can lead to future unnecessary expenditures of time, money, and manpower. Drilling through an electric conduit or cutting a water pipe are common mishaps that may be attributed to lost, missing, or out-of-date drawings and manuals.

Using PDF in the Bidding Process

The earlier part of this chapter identified the mountains of paper generated and moved about on a design and construction project. In this part of the chapter you see how an equivalent process can be managed digitally.

Electronic bidsets have seen some success in recent years. The US Army Corps of Engineers Engineering Research and Development Center runs an electronic plans and specifications service for the Corps and other agencies. The Corps has used PDF for the specifications for some time, but use CALS for the drawings. The Corps have been experimenting with PDF drawings and may eventually go to an exclusively PDF-based system.

Electronic Specifications

Specifications are composed of a series of word processing, spreadsheets, and scheduling files. The design firm develops a set of the various sections and divisions for projects. Each section or division is an individual file; the appropriate section and division files are collected and modifed for a specific project. A typical project often contains 50 – 100 word processing files. In addition, specifications usually have at least one spreadsheet and/or a schedule, and may also include scanned maps.

Both Acrobat 6 and 7 have the ability to combine all the electronic formats into one file in one step using the Create PDF from Multiple Files command. The files are selected and ordered, and then converted to PDF and combined into a single file.

About the CALS Format

CALS is a proprietary raster format similar to TIFF. Like other raster formats, enlarging a CALS drawing results in pixelation. As they are in a raster format, CALS drawings do not contain fonts, and consequently lack the ability to be searched. On the other hand, PDF is a vector format like CAD drawings, which can be zoomed without any distortion of the drawings, as vector drawings are based on equations.

Bid Sheets

Word and Excel have both been used to make electronic forms, although neither program was intended for forms development. Form users had to have access to the same programs to complete the forms. Word forms change appearance as they are filled out; Excel forms, although they are visually more stable, may also have problems in their use.

PDF forms are becoming the industry standard. Since the free Adobe Reader is one of the most widely distributed programs in the world, the problem of the user not having access to the proper program for completing the form is eliminated. Acrobat 7 has added the ability for the form to be e-mailed back to the originator automatically by clicking a button on the form that is preconfigured to send the FDF (form data format) file containing the form data back to the originator. The person completing the form cannot electronically save the completed form with Adobe Reader, although the form can be printed.

Viewing PDF Plans

There are numerous CAD programs available, and numerous viewers have been developed to make the CAD drawings available. It is not uncommon for one project to require three viewers to display drawings.

PDF is a logical solution to the problem of inappropriate viewers, by converting all drawing types to PDF and then viewing with either Acrobat or Adobe Reader. Although some CAD viewers have a few specialized advantages over PDF, Adobe is rapidly advancing Acrobat to meet the challenge. Bentley, the creator of MicroStation, has realized the advantages of one format and now users can make PDF files directly from MicroStation V-8 2004.

With the development of an ISO Standard PDF/E (PDF/Engineering), eventually all CAD programs will require PDF file-generating capabilities.

An Electronic Bidding Process

Figure 2.2 illustrates the electronic bidding process with one bidder. This is more like a design/build project. In this particular figure the Facilities Manager is the owner's representative or designer. An electronic PDF package is prepared from the various design programs. Security and digital signatures are added to the package to protect the content.

The Builder's Project Manager (PM) determines the various disciplines that need to work on the estimate and enables Adobe Reader's commenting capabilities for their review. Figure 2.2 does not include the process of receiving bids from subcontractors.

Once the estimate is finished, the PM completes the bid, digitally signs the document, and then electronically returns the information to the Facilities Manager for review by the project team. The Facilities Manager digitally signs the documents, and electronic copies are archived and forwarded to the Project Manager.

The electronic process saves a great deal of time, as well as the expense of printing, shipping, duplicating, collating, and reshipping the plans, specifications and contracts.

Summary

In this chapter we looked at two different processes for developing and administering a project workflow. In the first example, you saw how a traditional AEC project is developed and administered using a paper-based method. In the second example, a design/build workflow is illustrated, using an electronic PDF-based workflow.

Exercises

1. Using Figure 2.1, and the information outlined in this chapter, identify the processes in a traditional paper-based workflow.

2. Using Figure 2.1, determine areas in which savings could be realized by using an electronic workflow rather than a paper-based workflow.

3. Using Figure 2.2, examine a workflow that has been traditionally paper based, and determine ways to convert the workflow to an electronic format.

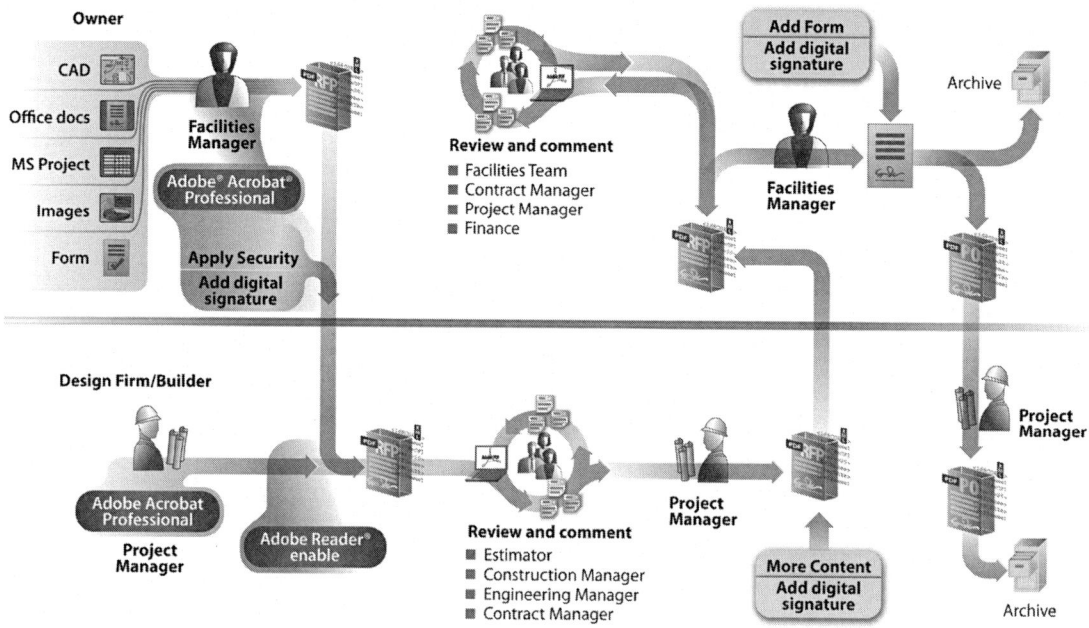

© 2004 Adobe Systems Incorporated. All rights reserved. Adobe and Acrobat is/are either [a] registered trademark[s] of Adobe Systems Incorporated in the United States and/or other countries.

Figure 2.2 An Acrobat-based AEC workflow

References

[1]   CSINet, *MasterFormat^{TM} 2004 Edition Numbers and Titles*, 7 July 2005. http://www.csinet.org/
s_csi/pdf.asp?TP=/s_csi/docs/9400/9361.pdf

3

Overview of Adobe Acrobat 7

Remember back in Fluids when the professor spent half a class deriving an equation? Remember how you kept saying to yourself "Just give me the equation, tell me when it works, and any problems I might run into."?

We are going to teach Acrobat the way you wanted Fluids taught. Acrobat is used by many different professionals for many different purposes. We are only going to teach you about one-half of Acrobat and when to use that one-half.

Acrobat allows you to personalize the program's layout according to your working style.

In this Chapter

In this chapter take a quick tour of the program and look at some of its features and tools, including:

- A list of program highlights in the latest version of Acrobat 7 Professional
- Comparison of different Acrobat 7 program versions
- How the program interface is displayed
- Program window components
- Accessing Help
- How to open, save, and close files

Program Highlights

Some of the highlights of the latest version of Acrobat Professional for Windows include:

- Create PDF documents from additional Microsoft Office programs in Windows, including Outlook, Project and Microsoft Access.
- Create PDF documents from engineering programs, including AutoCAD and Microsoft Visio.
- Mark up drawings using Measuring and Drawing Markups tools.
- Embed object-level data in Visio drawings that are converted to PDF.

- Create interactive forms with Adobe Designer 7©, an integrated program in Acrobat 7 Professional.
- Attach external files to a PDF document; the attachments can be extracted and saved as separate files in their native formats.
- Maintain customized collections of documents using the Organizer.
- Use the Managing Security Policies window to create and control security policies.
- Enable Commenting tools for Adobe Reader 7 users, and manage reviews using the Tracker window.
- Use the Print Production tools to create a PDF workflow for high-resolution output according to PDF/X publishing standards.
- Build document collections that comply with PDF/A archival standards.

Acrobat Versions

Acrobat is available in several versions. Table 3.1 outlines the differences in the features available in the alternate program versions. The table includes both Acrobat 7 Professional and Standard versions, as well as Adobe Reader 7, and Acrobat Elements©, a specialized version of the program with limited functionality for use in enterprise environments.

Table 3.1 Acrobat Versions and Capabilities

	Acrobat 7 Professional	Acrobat 7 Standard	Acrobat Elements	Adobe Reader 7
View, search, and print PDF files	X	X	X	X
Enable Adobe Reader 7 users to work with Commenting tools	X			
Manage document reviews	X	X		
Create PDF documents from applications that print	X	X	X	
Combine multiple documents into a PDF file	X	X		
Use content from Microsoft Office programs	X	X		
Use content from AutoCAD, Visio, and Microsoft Project	X			
Use password, certificate, and policy security	X	X	X	
Convert layers and object data in drawings	X			
Build intelligent forms using Adobe Designer	X			
Preflight high-end print jobs	X			

The Acrobat Interface

Acrobat Professional's program interface is composed of menus, toolbars, and specialized work areas or panes. There are multiple ways to access the same program's features and functions using some combination of menu commands, toolbar items, a right-click shortcut menu, and shortcut keys.

The default program window shows several components. You can configure the layout by displaying or collapsing panel groupings. As your proficiency with the program develops, you will likely find it simpler to hide or show different panes and toolbars.

Menus

An open PDF file is shown by name at the top of the Acrobat window. Acrobat contains a set of menus across the top of the program window (Figure 3.1). Many of the menu listings are similar to those of other programs. Acrobat contains several program-specific menu headings as well. If you install third-party plug-ins, you may find an additional menu heading added to the program.

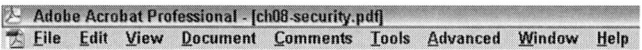

Figure 3.1 The program menu contains common and Acrobat-specific menu headings

Acrobat includes nine menu headings:

- **File.** Open existing files or generate new files, manage printing, document properties, as well as comment and form contents.
- **Edit.** Access common Edit commands such as Copy and Paste, as well as Search and Find commands and program preferences.
- **View.** Select toolbars, Task Buttons, and Navigation panes, as well as control the document view; access automatic features, such as scrolling and Read Aloud, and control placement aids, such as grid lines and rulers.
- **Document.** Select from a number of page functions, such as inserting and deleting pages; add page features such as headers and footers, backgrounds and watermarks; control access to a document using digital signatures and other security options; capture text using optical character recognition (OCR).
- **Comments.** Select different commenting tools, manage a review, start a new review, manage comments.
- **Tools.** Access the tools included on Acrobat's eight main toolbars.
- **Advanced.** Select advanced Acrobat features, such as the PDF Optimizer, Accessibility Checkers, batch processing, cataloging, and managing security policies.
- **Window.** Contains common window display features, as well as some Acrobat-specific features, such as using a split view, full screen view, or open the Clipboard Viewer.
- **Help.** Access commands for displaying both the How To help listings as well as the full Help files; includes online information, and Detect and Repair, used to correct program errors.

Toolbars and Task Buttons

Acrobat uses a large number of toolbars; there are also additional subtoolbars accessible through the main toolbars. Most tools are available through the View > Toolbars menu and the Tools menu.

> **TIP:** Save time opening and closing toolbars by using the toolbar well. Right-click a blank space on the toolbar area to open the list of toolbars and click to select a toolbar for display.

To refer to a name quickly, move your pointer over the vertical hatched lines at the left of a toolbar to display its name in a tooltip or pause your mouse over a tool to display the tool's name (Figure 3.2). For ease of use, you can show the tools' labels by choosing View > Toolbars > Show Button Labels, and then selecting the Default, All, or No button label options.

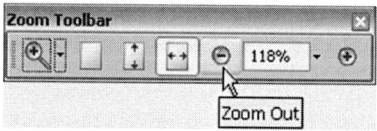

Figure 3.2 You can see both the toolbar's and individual tools' names in tooltips

Configure the toolbars' arrangement as desired by dragging toolbars off the toolbar area on the program window to float above the program, or drag them to redock with the other toolbars. Choose Lock Toolbars from the View > Toolbars menu, or by right-clicking the toolbar well to display the shortcut menu to lock the position of all docked toolbars. To reset the toolbars to the program default, choose Reset Toolbars from the menu.

The Task buttons are different from toolbars in that each contains a coordinating set of commands for performing a function, such as securing a document or creating a PDF document (Figure 3.3). Display a Task button with the toolbars by choosing View > Task buttons and selecting the appropriate Task button.

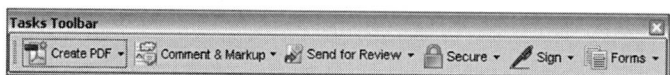

Figure 3.3 The Task buttons contain associated commands used for performing a type of task

Acrobat contains a set of six Task buttons:

- **Create PDF.** Generate PDF files using different methods, such as multiple documents or scans.
- **Comment & Markup.** Display different Commenting toolbars, manage comments.
- **Send for Review.** Choose commands for managing a review, starting a review, or opening the Tracker, the window used to control a review.
- **Secure.** Select commands for applying and managing security policies.
- **Sign.** Choose commands for adding secure signatures to a document.
- **Forms.** Choose commands for creating, editing, and converting forms, as well as other forms-related tasks.

Window Panes

Most of the program window is divided into three panes (Figure 3.4). The panes can be resized by dragging the vertical separator bars; only the How To window has a minimum allowable width.

The program window is comprised of these panes:

- **Navigation pane.** The tabs at the left side of the program window comprise the Navigation pane. Use the various tabs to manage and control document content.
- **Document pane.** The document displays in the central Document pane. Adjust the view of the document in the pane using the scrollbars.

- **How To window.** The How To window is shown at the right of the program window and displays the How To window, which contains links to information about performing common tasks, as well as the complete Help files. The same window on the program interface is used to perform searches and display results of Accessibility testing.

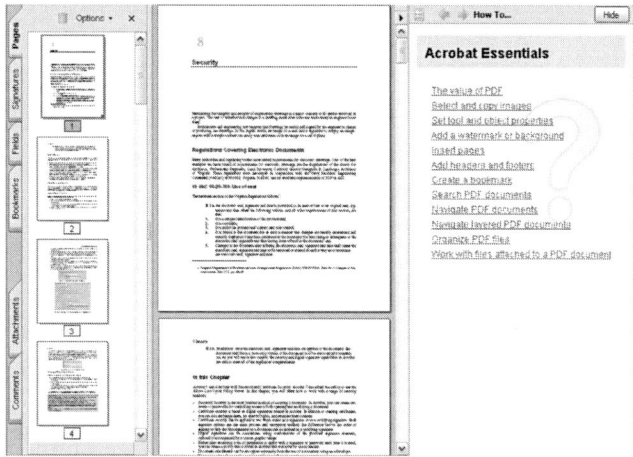

Figure 3.4 Adjust the width of the panes for ease of use

The Status Bar

The Status bar runs horizontally across the bottom of the program window (Figure 3.5). There are controls in the Status bar used to change layout views and navigate through a document numerically, as well as display information about the document:

- **Special status.** Any special features of the document are indicated by icons at the left of the Status bar.
- **Document view.** Click the Full-screen view icon ▢ to display only the document against a black background; click the Hide Toolbars icon ▢ to hide the program's menus and toolbars. When active, the icon toggles to a Show Toolbars icon.
- **Page size.** If you are displaying a document at full width or lower magnification, you can see the page's dimensions by moving your mouse into the lower left of the program window; the page size is shown in a tooltip (Figure 3.6). When the view is magnified and scrollbars are visible, the dimensions are shown on the horizontal toolbar.
- **Navigation Controls.** Use the controls on the Status bar below the Document pane to move between pages in the document; type a page number in the field to move numerically through a document. The field displays the number of the visible page and the total page count when the document is paginated.
- **View Buttons.** Acrobat captures the views you use during a session; use the two view buttons at the right of the Navigation Controls to move through the stored views. Click the Previous View (the left arrow button) to go to the last document page and magnification used in the Document pane; click the Next View (the right arrow button) to move to the next document page and magnification used in the Document pane.
- **Viewing Layout Controls.** The viewing layout options are indicated by a set of icons at the right of the Status bar. The views include single or continuous page, or book spreads.

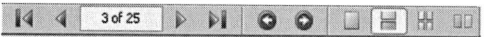

Figure 3.5 Control the document display using the Status bar features

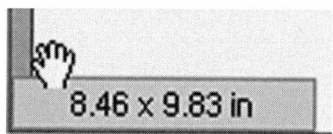

Figure 3.6 See the page's dimensions in a tooltip when the view is smaller than the Document pane

Using Help Features

The How To window displays by default when you open Acrobat. Each of Acrobat's Task buttons include a command to display help for that specific type of task. For instance, if you are learning how to comment on a document, you can choose How To Add Comments & Markups from the Comment & Markup task button's menu to display a list of topics in the How To window.

Navigate through the Help files using the icons at the top of the window. Directional arrows take you to previous and subsequent views; click the Printer icon to print the displayed topic; click the plus (+) and minus (−) icons to zoom in or zoom out of the Help window for easier reading.

To access the main Help files, choose Help > Complete Acrobat 7.0 Help or press the F1 key. The Help files open in a separate window. As in other programs, you can access help in three ways:

- **Contents.** The Contents listing is shown by default when the Help files open (Figure 3.7). Topics are arranged in a hierarchy from general to specific topics.
- **Search.** Type the search term in the Find topics containing field and click Search to display a list of topics containing your search term. Click a listing to display the content in the main pane with the search term highlighted. To hide the highlighting, click the Help window to deselect the highlights.
- **Index.** Type the first few letters of the term in the field at the top of the Index pane. Matching listings are displayed in the body of the window.

Opening and Saving Documents

You do not create blank PDF files in Acrobat and then add content within the program. Instead, you open PDF documents or create new ones in a number of ways. A selected file, such as a Word document, can be converted from its native DOC format to PDF using an automatic file conversion process.

The Create PDF Task button's commands let you generate PDF files using several methods, including from a single or multiple files, from a scanner, a Web page, or using a Snapshot, which is content captured from a document that is used to create a new file.

In addition to saving a PDF file in the usual way by using the Save command on the File menu, you can also save a PDF file in a wide range of additional file formats. In the Save dialog box, click the Save as type drop-down arrow and choose a file format option from the list (Figure 3.8). Once you make a selection, you can click the Settings button to configure most types of files before saving them.

Close Acrobat in the same way as any other program; when you reopen the program, the toolbars are in the same arrangement as you set before closing the program. The Navigation panes may or may not be displayed in the same way depending on whether an initial view is set for the document. Acrobat allows

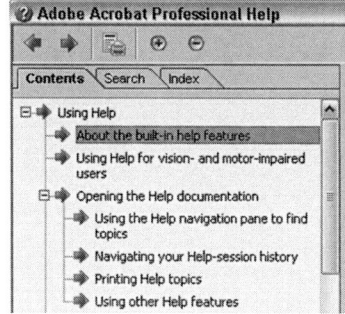

Figure 3.7 Use Help to find specific information and instruction

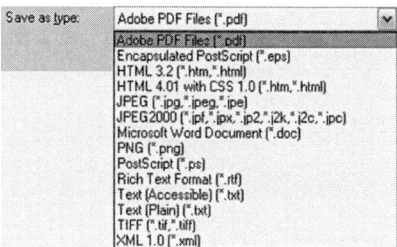

Figure 3.8 PDF files can be saved in a variety of formats

you to specify that one of several Navigation tabs is displayed when you open a file. Options include the Bookmarks, Pages, or Attachments pane. Display settings vary on a document by document basis.

Summary

In this chapter we looked at the Acrobat 7 Professional program, identifying some of its major features, and comparing different program versions. You saw how the program window is comprised of several components and how they interrelate. You learned how to open, close, and save files, and how to access Help in two different versions.

Exercises

1. Examine the contents of the program menus.

2. Open and close toolbars, rearrange, and lock a toolbar configuration.

3. Open the How To window using the How To options from the Task Buttons' menus, examine the options listed.

4

Creating Package Components from Documents

The key to producing PDF files in a range of programs is a *.joboptions* file. The .joboptions file format is used to apply a preconfigured group of settings to a file for conversion to PDF. You can create PDF files through a number of programs that include a PDFMaker macro, via Acrobat Distiller, or by using the Adobe PDF printer driver. Use the commands installed with the macro to configure conversion settings and control PDF document generation. Acrobat 7 includes a fantastic new feature for engineering management – you can now archive a project's emails in an easy-to-use, searchable format.

In this Chapter

Efficiently creating a PDF file from a source document is important to project development. You can create a PDF file using a variety of methods, and a range of preconfigured settings. In this chapter you learn how to:

- access a PDFMaker macro
- apply settings – in this chapter, a Word document is used as an example
- choose options based on the intended use of the document
- use the Adobe PDF printer driver
- work with Acrobat Distiller
- understand common conversion options and create custom conversion settings.

Creating PDF Files

As part of the program's installation process, Acrobat includes components in many Microsoft programs called *PDFMakers*. The PDFMaker is a macro designed to install both a toolbar and a menu used to create PDF files in the program into which it is installed, and then distribute the PDF file in a number of ways. The configuration options provided by the PDFMakers are both specific to the program within which the PDFMaker is installed, as well as containing a number of generic options that are available to all PDFMakers. The contents of an example Adobe PDF menu are shown in Figure 4.1.

In Windows, PDFMakers are installed in Word, Excel, PowerPoint, Publisher XP and 2003, and Outlook as part of the Acrobat Standard program installation; Acrobat Professional includes PDFMakers for AutoCAD, MS Project and Visio. The first time you open the program after installing Acrobat, you may see a message regarding enabling macros – choose to always trust macros from Adobe Systems and enable the macro to allow the PDFMaker to load.

In earlier versions of the program, you had to reinstall Acrobat after installing a program in order to attach the PDFMaker to the program. In the current version of Acrobat, use the Help > Detect and Repair command; Acrobat automatically installs the appropriate PDFMaker.

Regardless of the method used to create a PDF document, the same preconfigured conversion options can be used. These options are stored in the Acrobat program installation folder as .joboptions files; you can create and save customized settings, or import external settings.

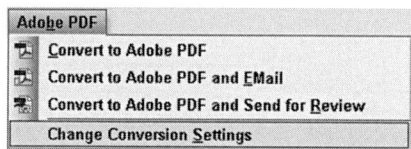

Figure 4.1 Word's PDFMaker contains four commands

Streamlining File Conversions

Regardless of the method you use to convert one or more files, keep these simple tips in mind to save time in the conversion process:
- Check spelling and grammar before conversion.
- Select the conversion option that is right for the job. If you intend a document to be used online, for example, you do not need high-quality print resolution.
- When converting a large number of documents, convert and test one document first.
- Check comments and links for accuracy in the source document before converting to PDF.
- Decide on the naming convention you will use. By default, Acrobat uses the source file's name as the PDF file's name as well.
- Deselect the conversion setting that displays the file automatically in Acrobat after conversion if you are converting a number of files to save time.
- If your work is for the government, designed to meet specific legal requirements, or for an ISO-certified company, consider using the PDF/A archival draft standards for all conversions.

Using the PDFMaker in Word

The most common program used in conjunction with Adobe Acrobat is Microsoft Word. The PDFMaker installed into Word generates a PDF document using the default *Standard* settings conversion option. Settings selected in a PDFMaker remain until they are reset; settings you choose in Word are also used if you convert a document to PDF from within Acrobat. To convert a file to PDF in Word, follow these steps:

1. Click Convert to PDF on the PDFMaker 7.0 toolbar 📄 or choose Adobe PDF > Convert to Adobe PDF.

2. In the Save As dialog box, the PDF file is shown using the original document's name and file location; choose an alternate name or storage location if required.

3. Click Save to close the dialog box and convert the file.

Viewing Conversion Settings

Customize the conversion settings or program-specific options by choosing Adobe PDF > Change Conversion Settings to open the four-tab Acrobat PDFMaker dialog box (Figure 4.2). Click the Conversion Settings drop-down arrow to choose a different conversion settings option; click Restore Defaults to revert to the program's default settings. All PDFMakers include Settings and Security tabs. Figure 4.2 shows the Acrobat PDFMaker defaults in Word.

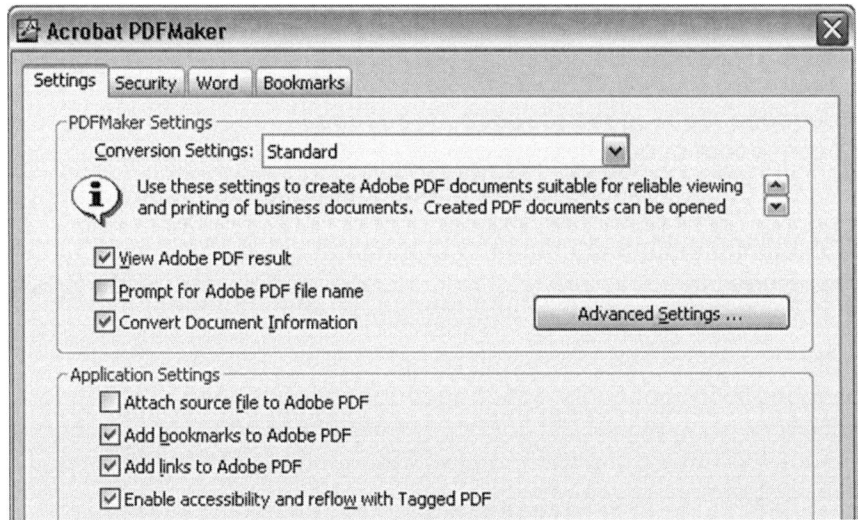

Figure 4.2 Choose conversion options before generating a PDF file

Conversion Options

Acrobat uses six types of PDF conversion, based either on intended use or its adherence to a standard:

- **Standard.** Standard settings are the default used for most business document conversion. These settings use a printing resolution of 600 dpi, but downsample graphics in an efficient manner that saves file space.
- **High Quality.** The High Quality settings use a printing resolution of 2400 dpi, and include a limited amount of information about the document's fonts.
- **Press.** Press settings are used for high-end print production, such as image setters. Output uses high resolution; the files contain all coded information about the document's fonts.
- **Smallest File Size.** Use the Smallest File Size option for documents intended for distribution by email, for onscreen use, or on Web pages. To achieve a small file size, fonts are not embedded, and images are compressed and resolution decreased.
- **PDF/A.** The PDF/A archival standard is used when documents are intended for long-term storage and use.
- **PDF/X.** Acrobat provides four PDF/X standards versions; use the standards for high-resolution print production. PDFMakers cannot produce a standards-compliant PDF/X document – you must use Acrobat Distiller or generate the compliance within Acrobat.

Common PDFMaker Options

Some of the PDFMaker options are consistent throughout all programs that use a PDFMaker macro, including:

- **View Adobe PDF result.** Deselect this option, chosen by default, if you do not need to work with the document in Acrobat.
- **Prompt for Adobe PDF file name.** Deselect this option if you prefer to use the same name as the source document, saving a step in the conversion process.
- **Convert Document Information.** Leave the document information option selected – including the data in the converted document adds little to the file size and provides information for manipulation in Acrobat.
- **Application Settings.** The Application Settings options change depending on the program you are working in. Select or deselect the options based on your intended use of the document.
  In Word, choose from these options (the default choices are shown in Figure 4.2):
    - **Attach source file to Adobe PDF.** The original document is appended to the PDF document and can be manipulated in Acrobat.
    - **Add bookmarks to Adobe PDF.** Styles or headings can be used as the basis for a set of bookmarks added to the PDF document in Acrobat.
    - **Add Links to Adobe PDF.** Links are automatically converted to PDF links.
    - **Enable accessibility and reflow with Tagged PDF.** In addition to using tags for viewing and accessibility purposes, you can also generate bookmarks or control content in Acrobat.

Security Settings

You can add password protection to a document via the Acrobat PDFMaker dialog box. Add passwords to a file only if you intend to distribute it – if you plan to work with it further in Acrobat, wait until the file is edited before adding security to save time inputting passwords each time you open the file in Acrobat. Refer to Chapter 10 for information on securing a document.

Word-specific Settings

Depending on the program in which it is installed, an Acrobat PDFMaker includes a program-specific tab containing configuration options. Word includes such a tab on which you can choose to convert comments, cross-references or tables of contents, and footnotes or endnotes (Figure 4.3). The latter two options are selected by default.

If you select to convert comments, you can specify the comments to convert based on the reviewer, and choose the color of the comment icon used in the PDF file, as well as whether or not the comments are open in the PDF file.

Generating Bookmarks

Bookmarks are one of the most common methods of navigating a PDF file. You can use the Bookmarks option in the Word PDFMaker to generate a set of bookmarks automatically based either on the document's styles or headings. Click the Bookmarks tab in the Acrobat PDFMaker dialog box to display the conversion options. The default is to convert headings, as shown in Figure 4.4.

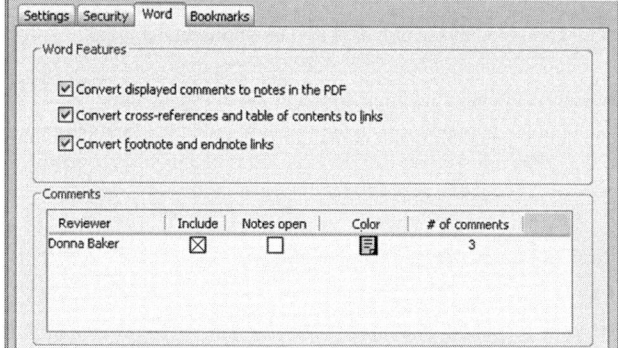

Figure 4.3 Choose Word-specific options to include in the conversion

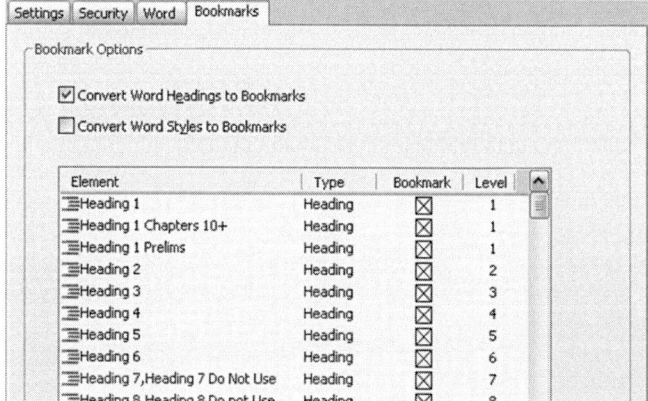

Figure 4.4 Select a method for creating bookmarks from your Word document.

Note: Read about bookmarks and other forms of navigation in Chapter 9.

Quick File Conversions

Commonly, you use the same settings for converting files to PDF. Rather than opening a document in its source program or converting it via Acrobat, you can use the shortcut menu. Open Windows Explorer, and right-click the file to be converted to display the shortcut menu. After PDFMakers are installed, the convert to PDF, convert and email, and convert and send for review options are included in the shortcut menu (Figure 4.5).

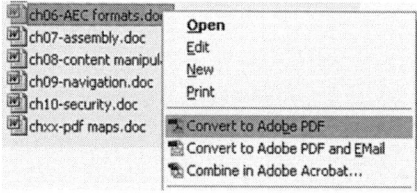

Figure 4.5 Use the shortcut options to convert files quickly

Producing PDF Files in Other Office Programs

Acrobat PDFMaker menus and toolbars are added to other Office programs. All PDFMaker macros include Settings and Security tabs in their Acrobat PDFMaker dialog boxes. Additional configuration options in both the Adobe PDF menu and the PDFMaker dialog box vary according to program, as listed in Table 4.1.

Printing with the Adobe PDF Printer

The Adobe PDF printer driver is installed in your system as part of the Acrobat 7 installation process. The Adobe PDF printer allows you to convert any document that you can print to a PDF file. To use the printer driver, follow these steps:

1. Open the document you want to convert, and choose File > Print to open the Print dialog box.
2. Click the drop-down arrow and choose Adobe PDF from the printer list.
3. Click OK to process the file.
4. Choose a name and storage location and click Save. The document is saved to PDF rather than sent to your printer.

You can modify options and preferences for the Adobe PDF printer driver. On the Print dialog, click Options to open the Print dialog box (Figure 4.6). In the Print options, you can choose to include several components along with the document, such as XML tags, field codes, and document properties. If you are printing a form, you can choose to print the form data only. Click OK to close the Print options dialog box and return to the main Print dialog box.

Click Properties on the main Print dialog box to select or modify the .joboptions that will be applied to the document (Figure 4.7). In the Adobe PDF Document Settings dialog box, select a conversion setting, security, output folder, and page size.

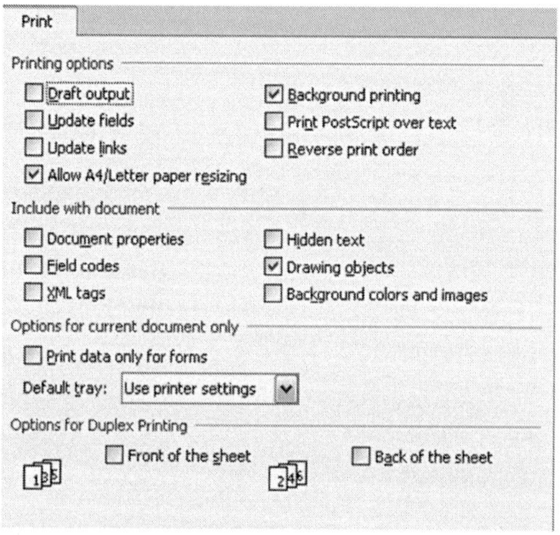

Figure 4.6 Select Adobe PDF printer driver options

Table 4.1 PDFMaker macro functions in Office programs

Program	PDFMaker Functionality
Excel	The Adobe PDF menu includes an option to convert an entire workbook; the default is to convert the visible worksheet. In the PDFMaker dialog box, in addition to the Application Settings options used by the Word PDFMaker, you can choose Fit a worksheet to a single page to rescale the spreadsheet contents; you can convert links, bookmarks, and comments as well
Access	The Adobe PDF menu includes an option to convert multiple reports to a single PDF document to combine an Access project's reports. The Access PDFMaker's Application settings allow you to include bookmarks or attach source files.
Project	In the Project PDFMaker dialog box, choose options for attaching the source file to the PDF and making the project fit on a single page. PDF conversion includes only the currently selected view. Test the converted Project file, as some views may not be compressed to a single page.
PowerPoint	The PDFMaker in PowerPoint has several configuration options. Options include settings for exporting slide transitions, text animations, or defining the page's layout using the presentation's print settings. You can convert comments, and bookmarks, or add document tags.
Publisher	Unlike other programs' PDFMakers that use the Standard settings as the default conversion option, Publisher's defaults to the Print settings. Publisher documents are often used for high-end printing, and allow you to choose print-specific options including spot colors, crop marks, and transparency. Converted Publisher documents can also contain bookmarks, tags, and comments.

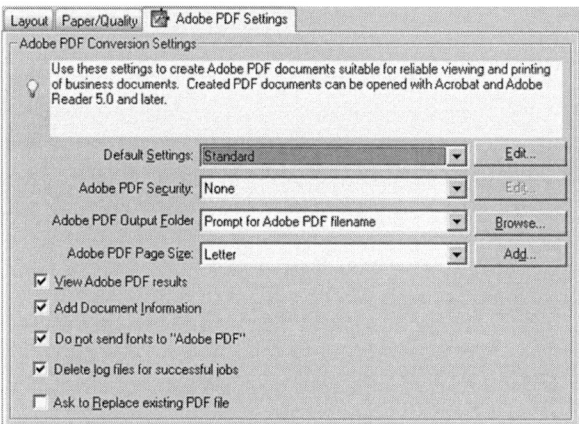

Figure 4.7 Modify properties for the Adobe PDF printer driver

Choose other conversion settings – the defaults are shown in Figure 4.7:

- The View Adobe PDF results opens the file automatically in Acrobat after it is generated. Deselect this option if you are printing to file for use at a later time.
- Add Document Information stores font and structural information about the file used in procedures such as indexing or searching or modifying content.

- The Do not send fonts to "Adobe PDF" option is selected by default, producing a smaller file size without including font information. If you intend to use the file in Acrobat Distiller, deselect this option as Distiller requires embedded fonts.
- Delete log files for successful jobs, chosen by default, removes logs of completed jobs – this setting is selected by default.
- Ask to Replace existing PDF file, deselected by default, will ask if you want to replace an existing file. Depending on your favored workflow, either leave the default or use the file replacement prompt.

Archiving Email in Outlook

A common practice requires you to archive and store all email messages for a project, which can result in unwieldy binders of printed messages, files stored on disk, and material lost or misfiled. Using Acrobat, you can archive Outlook email messages, add them to existing PDF files, or index and search them from within Acrobat. Messages can be archived as single PDF files or appended to existing files. You can either create PDF files from single emails or entire folders. For archiving messages in Outlook, you can choose among several options.

To create a PDF file from a message or messages, select the message(s) in Outlook, and click Create PDF From Selected Messages on the Outlook Adobe PDF toolbar 📧 or select an option from the Adobe PDF menu (Figure 4.8).

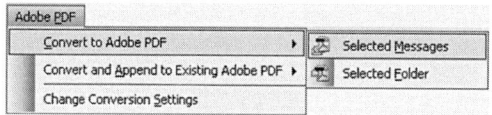

Figure 4.8 Convert Outlook email messages using the PDFMaker

To create a PDF document from the contents of an Outlook folder, select the folder and click Convert selected folder to PDF 📧, or choose the command from the PDFMaker menu.

To add files to existing archives, select the file you want to add from the Outlook messages and click Convert and Append to Existing Adobe PDF 📧; in the resulting dialog box, select the file to which you want to append the selected file or folder.

When you open the archive in Acrobat, you see bookmarks added to the file which arrange the content by date, sender, and subject (Figure 4.9). If you have added files from several folders, you also see a Personal Folders (or similarly-named) listing.

Configuring an Outlook Archive's Appearance

Choose Adobe PDF > Change Conversion Settings to open the Acrobat PDFMaker dialog box. Make these selections:
- Choose the appropriate program version from the Compatibility drop-down menu, which defaults to Acrobat 5.
- Choose an option for attaching the original email message; specify the page size and margins for the converted email messages. Click OK to close the dialog box.
- The settings you specify in the dialog box will remain until you reopen the dialog box and change them.

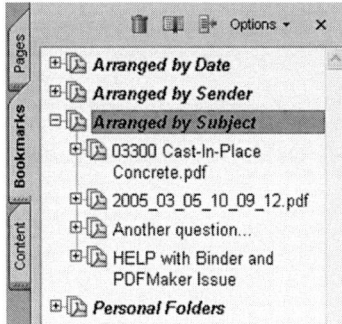

Figure 4.9 The Outlook PDFMaker adds bookmarks automatically

Preparing a Layered Visio Document

Microsoft Visio includes a PDFMaker that can be used to convert a layered drawing to PDF while allowing you to customize the layer conversion options. Visio's default is to flatten a document, but you can also maintain all layers or specified layers in the converted PDF file.

Like other PDFMaker macros, you can select from basic conversion options in the Adobe PDFMaker's menu, or choose an option to convert all layers in the drawing.

Before converting a Visio file to PDF, establish appropriate settings in Visio's Layer Properties dialog box as the settings chosen in the Layer Properties dialog box are used in the PDF conversion. Check settings such as layer visibility, and print status (Figure 4.10).

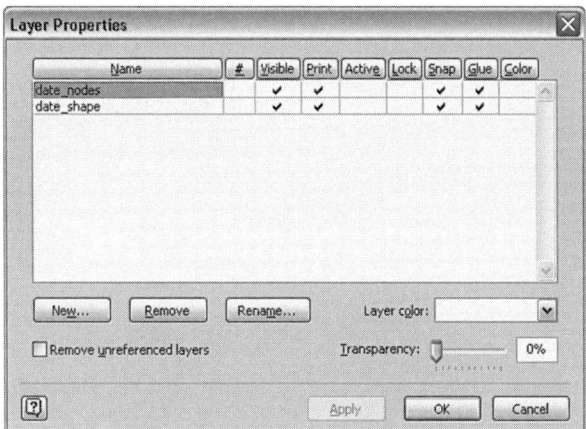

Figure 4.10 Set layer characteristics in Visio's Layer Properties dialog box

Converting a Visio File to PDF

Once the file's layers are configured as desired, choose Adobe PDF > Change Conversion Settings to open the Acrobat PDFMaker dialog box and check the conversion options. Select basic conversion and

Application Settings as shown in Figure 4.11. Like other PDFMakers, you can select common options for viewing, attaching the source file, and adding bookmarks or links.

In addition, there are several Visio-specific options on the Settings dialog box. These include choices for converting custom properties to object data, managing the layers using a default setting to always flatten the layers, and whether or not to display the Layers pane in Acrobat. When the settings are chosen, click OK to close the dialog box.

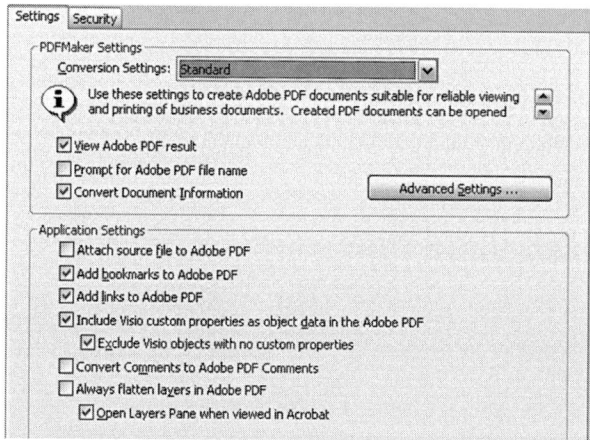

Figure 4.11 Choose basic application and PDFMaker settings

Converting a Visio file to PDF uses a sequence of dialog boxes. To convert the document to PDF, follow these steps:

1. Click the Convert to Adobe PDF icon 📄 on the PDFMaker toolbar, or choose Adobe PDF > Convert to PDF to start the process.

2. The first pane of the dialog box that opens describes the use of custom properties. Click the Include Custom Properties check box to convert custom properties added to shapes in the Visio file to Object Data in the converted PDF file, which is viewable in Acrobat using the Object Data tool. Click Continue to display the next pane of the dialog box.

3. Select a layering option for the document – you can flatten all layers, retain all layers, or retain some layers (Figure 4.12). If you modified the Application Settings on the Acrobat PDFMaker dialog box, the Flatten all Layers option is selected by default. For assistance, click Help to display a popup window describing the layering options. Click Continue.

4. As the option to retain some layers was selected in the example, the next pane of the dialog box shows the list of available layers (Figure 4.13). Click a layer in the Layers in Visio Drawing column and then click Add Layer to add the layer to the Layers in PDF column. Layers listed in the Layers in PDF column are those that are exported in the final PDF file.

5. The final pane of the dialog box lists your choices; click Convert to PDF to create the output. You can click Do not show this step again to bypass the confirmation dialog box and convert the document automatically, saving one step in future conversions.

6. Name the file and choose the storage location. Click Save to generate the PDF file.

Note: You can also create layer sets, or save the settings, which are then displayed by default the next time you convert a Visio document. These methods are described later in the chapter in the Converting an AutoCAD File section.

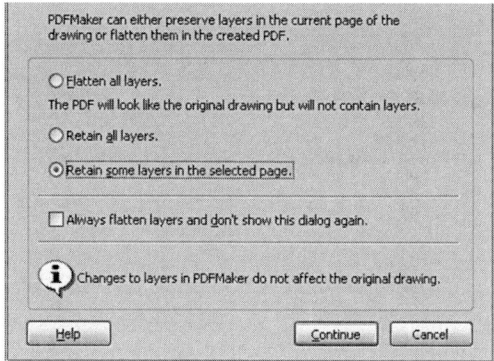

Figure 4.12 Select a layer display option

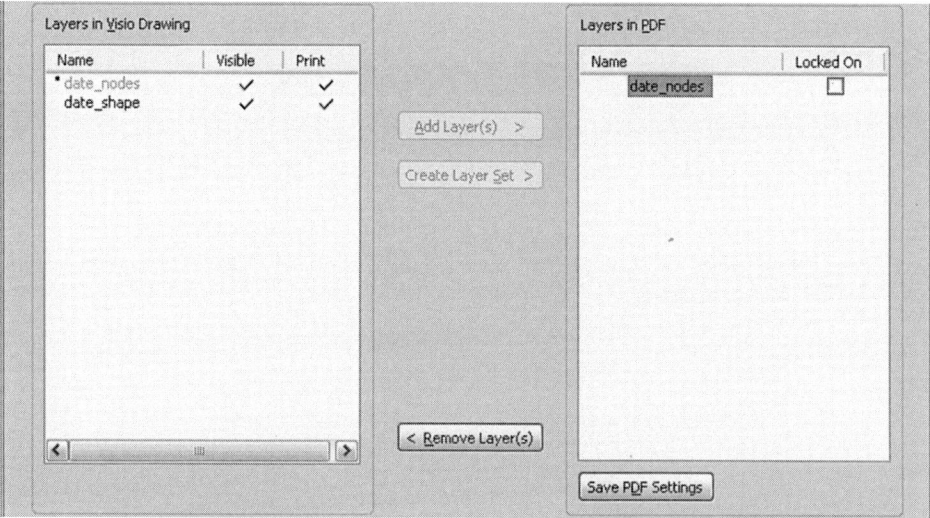

Figure 4.13 Choose the layers to preserve in the exported PDF document

Organizing Layers

Whether you convert a Visio or AutoCAD drawing to PDF, you manipulate the layers in the same way in the PDFMaker when converting a subset of the drawing to a PDF file:

- Layers that have been added to the Layers in PDF column are grayed out in the Layers in Drawing column; click a layer in the Layers in PDF column and click Remove Layer to delete it from the list of layers for export. The layer is then reactivated in the Layers in Drawing column.
- Activate the text and rename a layer in the Layers in Drawing column; the PDF file lists the layers by name in the Layers pane.
- You can create subfolders to organize groups of layers. Select a layer in the Layers in Drawing column. Click Create Layer Set to add a folder to the Layers in PDF column and nest the selected layer. Type a name for the layer set.
- Organize layers by dragging in the Layers in PDF column – drag up or down to reset the order; drag a layer below the name of a layer set to add the layer to the layer set.

Using Acrobat Distiller

Acrobat Distiller is a separate program installed as part of the Acrobat 7 installation process. Distiller's interface works like a dialog box. In Distiller, you choose the same conversion settings as those used in the PDFMakers.

Distiller is used to convert a range of file formats such as .ps, .eps or .prn files to PDF. To distill a file, choose File > Open to select the file for conversion; click Open to close the dialog box. As the file is processed, Distiller displays details and a progress bar.

Continue converting other files as required or close Distiller. You can manage the files in your Distiller session from the program's dialog box (Figure 4.14). Right-click a processed file to open the shortcut menu, where you can manage history and log files. The results listed in the Distiller dialog box are maintained for a single session; if you close and reopen Distiller, the list is cleared.

Errors are shown in the Distiller history list, indicated by an error icon, and point to a log file describing the errors. In Figure 4.14, an InDesign CS file produces an error as it is not a format compatible with Distiller.

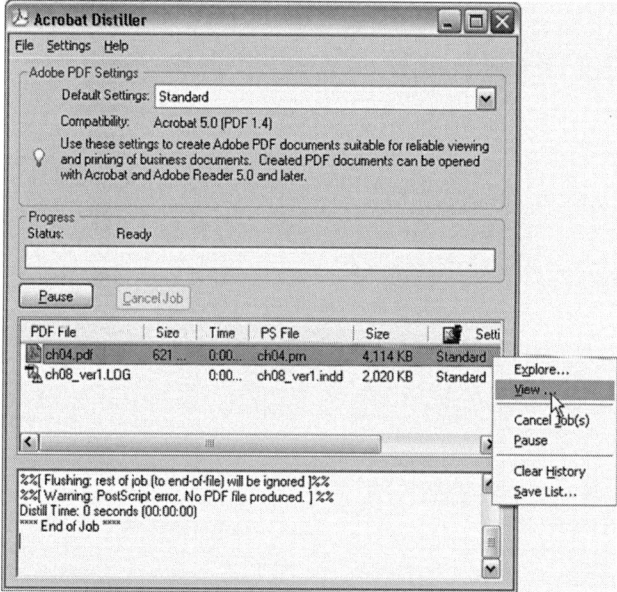

Figure 4.14 Distiller displays details of file conversions.

Creating Custom Conversion Settings

Acrobat Distiller, Acrobat, and the Acrobat PDFMakers share conversion settings, and generate the same output using the same selected conversion setting. You can use the default options, modify existing options, or create custom conversion settings. It is usually simpler to modify an existing option that is close to your requirements, and then save it as a custom .joboptions file, as in these steps working from Distiller:

1. In Distiller, choose one of the default settings to serve as the basis for your custom settings – this example uses the Standard settings as the basis for custom .joboptions file.

2. Choose Settings > Edit Adobe PDF Settings to open the Standard – Adobe PDF Settings dialog box (Figure 4.15).

3. The dialog box lists headings in the left column; click a heading to display its set of options in the right pane of the dialog box. Both the General and Standard headings show the same General options.

4. Adjust the settings as required for your custom file, described in following sections.

5. Click Save As; type a name for the custom settings in the resulting dialog box and click Save.

Note: Custom settings files can be shared like other files for uniform conversion results across a project. You can email a file to another user, who can include the file with their other .joboptions files.

General Settings

The General settings, shown in Figure 4.15, define basic characteristics for the conversion. The default compatibility version is Acrobat 5.0 (PDF 1.4); specify any compatibility from Acrobat 3 to Acrobat 7, as used in the example. The older the PDF version selected, the larger the pool of users that can view the document. However, the older the version, the fewer options for fonts, color, and security. In engineering use, Acrobat 5 is suggested as a minimum, as earlier versions have limited use with CAD drawings. Other common settings include:

- **Object level compression.** Compression of objects combines small objects into compressible content. Choose from Off, Tags only, or Maximum options. If you leave the compression set to Off, structure and tag information is usable in Acrobat 5; choosing either of the other options allows information to be usable in only Acrobat 6 and 7.
- **Resolution.** The default resolution is set at 600 dpi; you can emulate the resolution of a printer for PostScript files by adjusting the figure. Higher resolution usually produces higher quality files, more steps in a gradient or blend, and larger file size.
- **Page Range.** A common error is to specify a page range. Unless the settings are designed for a one-time use with a specific document, reusing the settings can yield erroneous results, as only the specified pages are converted.
- **Embed Thumbnails.** Enabling Embed Thumbnails adds to the file size unnecessarily; do not select the option unless the output is designed for older program versions.

Image Settings

There are several common image settings that may be need to be modified if you are working with specific types of image (Figure 4.16). Common changes are made to downsampling, compression and quality; you can also define a policy to process images consistently:

- **Downsample.** Pixels in images with higher than the specified resolution are combined to decrease resolution in a process known as *downsampling*. For images where the user is likely to zoom in to a high magnification, such as a map, a high resolution usually produces more legible images.
- **Compression/Image Quality.** Select individualized settings for color, grayscale, and monochromatic images. For monochromatic images, you can select anti-aliasing to prevent jagged edges.
- **Policy.** Click the Policy button to display a dialog box to specify whether to ignore, warn, or cancel a job when processing an image that falls below specified resolution.

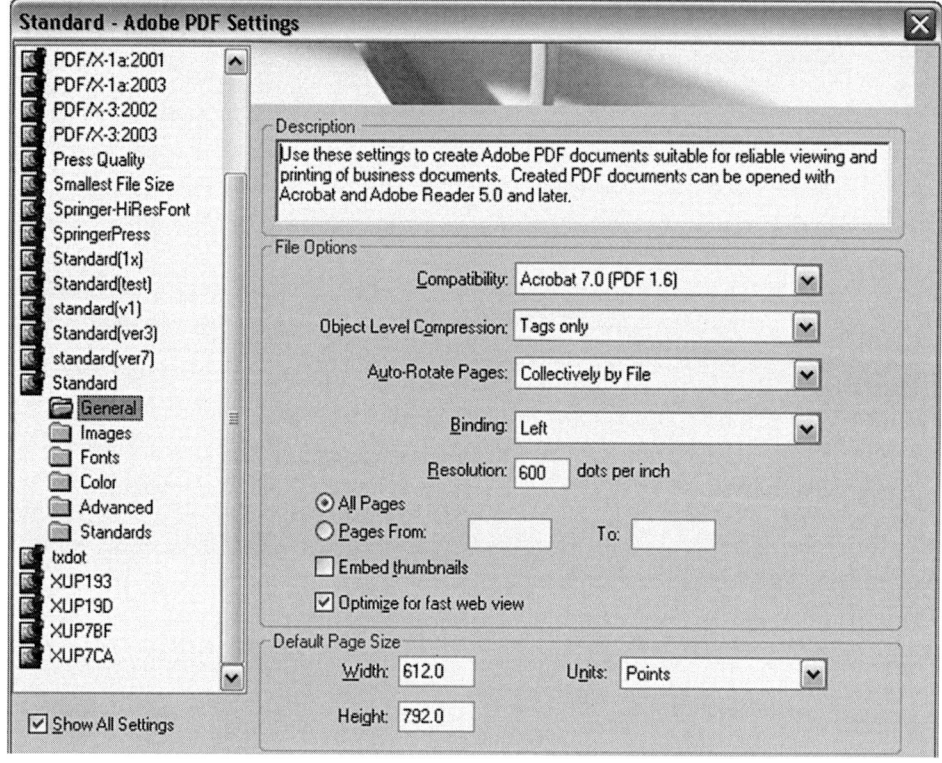

Figure 4.15 Configure and save custom .joboptions files

Tip: Choose File > Preferences to open the Preferences and select options for startup, output, and log files.

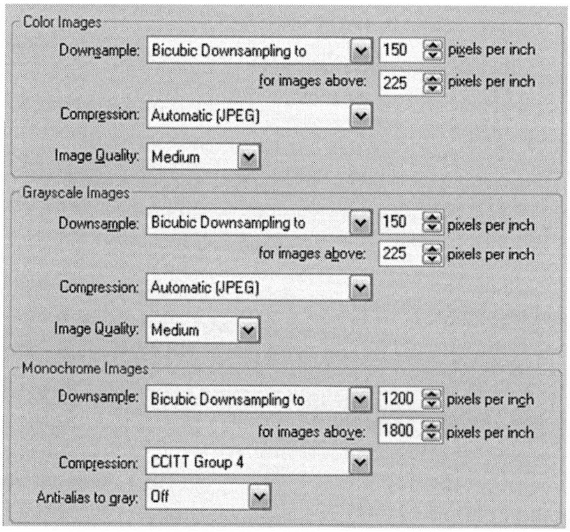

Figure 4.16 Choose custom image settings and conversion policies

Font Settings

On the Fonts panel, specify how fonts should be embedded and subset when the percentage of characters used falls below a specified value (Figure 4.17). When using a specific font for visual effect, be sure to embed the font; if you intend to work with the text in Acrobat, use a high subsetting level, or leave the default at 100% subsetting, which means information on all characters in the font are stored in the PDF file.

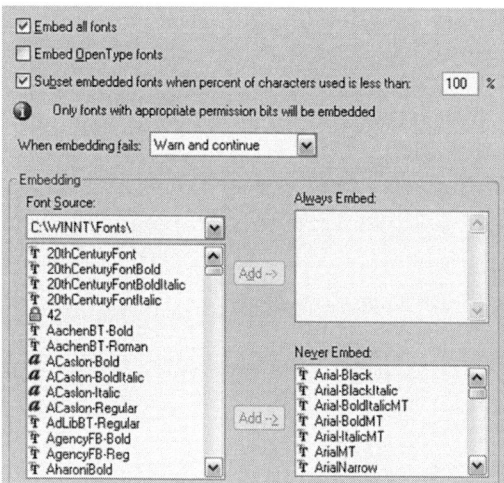

Figure 4.17 Choose font embedding and subsetting characteristics

Color Settings

Choose settings that correspond with files used in your source applications, such as Adobe Photoshop or Illustrator in the Color Settings. The available options depend on the color setting selected (Figure 4.18).

Advanced Settings

The Advanced Settings panel displays settings for Document Structuring Conventions, describing how the conversion from PostScript to PDF is carried out (Figure 4.19). The default settings are usually sufficient. Some changes may be useful, depending on the file contents, such as:

- **Convert gradients to smooth shades.** Used to convert gradients from a range of programs; use this option to produce smoother output without necessarily having to increase the resolution of the PDF file.
- **Convert smooth lines to curves option.** PDF files are smaller and redraw faster in a CAD drawing when you use this option to reduce the number of control points in a curve.
- **Save Adobe PDF Settings Inside PDF File.** Embeds the settings file used to create the PDF file in the Attachments tab.
- **Save original JPEG images in PDF if possible.** Use this option when the document for conversion contains numerous JPEG images to process the source files without recompressing the JPEG images.

Custom .joboptions File Tips

Here are a few tips to make creating custom .joboptions files simpler and more useful:
- To change the base conversion settings when creating a custom file, click the Show all Settings checkbox at the lower left of the left column on the Adobe PDF Settings dialog box to open a list of Default Settings; double-click another default option to display its headings, and proceed with the customization.
- Use the default storage location on your hard drive when saving a custom settings file to include the custom file within the Acrobat installation folders.
- Name the .joboptions files using meaningful names. Acrobat names custom settings using a default nomenclature of the conversion setting type followed by a number, such as standard(5).joboptions.

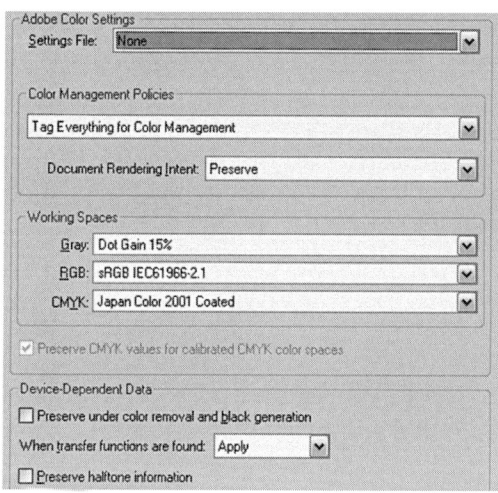

Figure 4.18 Correlate color settings with source application settings

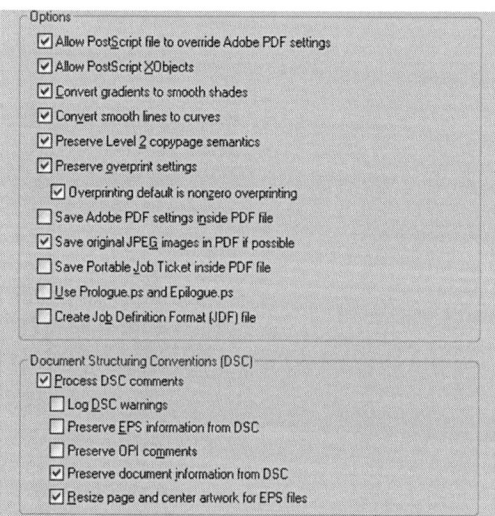

Figure 4.19 Modify options for converting specific types of content

Standards Settings

When your document must comply with established Standards, choose an option from the Standards panel to test the document against the standard's requirements before conversion. Figure 4.20 shows the PDF/A archival standard's settings. Click the Help icons next to an option to display a tooltip describing default values for the option, as shown in the figure.

Summary

In this chapter we looked at the method Acrobat uses to convert documents to PDF, using a Word document as an example. You saw how the PDFMakers, Acrobat Distiller, and Acrobat share a group of common conversion settings, called .joboption files. These are installed as part of the Acrobat 7 installation process and stored on your hard drive.

You saw how a PDF document is created from Visio, and how to manipulate the layers in the file. You learned how to print a file using the Adobe PDF printer driver, and how to convert a file using Acrobat Distiller. In addition, you saw how to examine conversion options, and also how to choose common settings for creating a custom .joboption file.

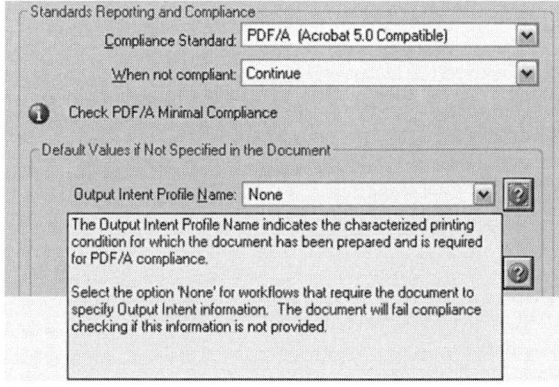

Figure 4.20 Choose settings to generate and test standards-compliant documents

Exercises

1. Using the method outlined in the chapter, convert a sample Word document; depending on your available programs, convert other sample documents using various programs' PDFMakers.

2. Convert and examine an Outlook PDF file. Note the bookmark structure added in the PDF file.

3. Convert a sample file using Acrobat Distiller. Note the conversion process, and look at the log file generated by the program.

4. Create and save a custom .joboptions file, either through Distiller (as used in the chapter), or via a program's PDFMaker settings dialog box.

Copyright Notice: Many elements of this project have been modified due to copyright and liability issues, and should not be used in other projects. They are only being used to teach the processes in the context of this book.

Project

The goal in this project is to create three versions of the project's cover page, using two different conversion methods. Use the source file in the **ch04_project** folder.

Task 1: Converting a file from an Office product

1. Open the Word file 0_Cover.doc in the ch04_project folder.
2. Convert the file by clicking on the Convert to PDF icon using the PDFMaker's default Standard conversion option. Save the file to your hard drive for use in Chapter 5.

Task 2: Converting a file with custom conversion settings

1. Change the conversion settings for the PDFMaker by choosing Adobe PDF > Change Conversion Settings. In the Conversion Settings dialog box, choose PDF/A Draft.
2. Convert the **0_Cover.doc** file to PDF using the custom setting.

Task 3: Comparing file properties

1. Open each file in Acrobat.
2. For each version, choose File>Document Properties>Description. Compare File Size, Tagging and Fast Web View. You find the files using the PDF/A standard are much larger and do not include Fast Web View.

Task 4: Converting with the AdobePDF printer driver

1. Using the same Word file, choose File> Print and select the Adobe PDF printer driver from the printer drop-down menu.
2. Print the file.

5

Creating Other Package Components

Rather than opening programs and converting files to PDF from their native applications, you can instead create package components as PDF files within Acrobat in a number of ways. The method you use depends both on your intended use of the files and your preferred workflow.

In this Chapter

You can create package components as PDF files from other source documents. In this chapter you learn how to:

- Create PDF files from single or multiple documents
- Create PDF files from a scan, and capture and test the content
- Create PDF files from a Web page, either through Acrobat or Internet Explorer
- Create a PDF file from captured content, either in the form of a snapshot or from clipboard contents.

Converting a Single Document

Acrobat 6 and 7 include a Create PDF task button, which contains the options for creating a PDF file within Acrobat. Click the task button to display its menu, shown in Figure 5.1, or choose File > Create PDF menu. For assistance, choose How to...Create PDF from the Task button's menu to open common file creation topics in the Search pane at the right of the program window.

To create a PDF from a single file, choose From File from the Create PDF task button's menu. Select the file you want to convert in the Open dialog box and click Open. The file is opened in its source program and converted to PDF. Acrobat uses a PDFMaker with its current settings in applicable programs; if the file's native program does not have a PDFMaker, the file is printed using the Adobe PDF Converter. After conversion, the document is opened in Acrobat, and named using the source file's name. Choose File > Save to save the file as a PDF.

Not all files can be converted to a PDF from within Acrobat; allowed formats are listed in Table 5.1.

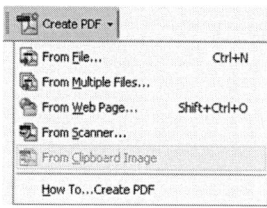

Figure 5.1 Use the task button's menu to choose operations

Table 5.1 File formats converted to PDF via Acrobat

AutoDesk AutoCAD	AutoDesk Inventor	BMP	EPS	GIF	HTML
JDF Job Definition	JPEG2000	Microsoft Access	Microsoft Excel	Microsoft PowerPoint	Microsoft Project
Microsoft Publisher	Microsoft Visio	Microsoft Word	PCX	PNG	PostScript
RTF	Text	TIFF			

Where to Convert a Document

Many documents can be converted either from their source programs or from within Acrobat. The option you choose depends on your workflow. If you are presently working in the source program, generate the PDF document; if you are working in Acrobat, generate the document from Acrobat.

The exception is when you have to modify settings. If you have forgotten the last settings you applied in a PDFMaker, work from the source program so you can check the settings and choose alternate options if required.

Converting Multiple Documents

It is not necessary to convert a number of documents to PDF in their native programs and then combine them in Acrobat. Instead, you can convert and combine a combination of PDF and other document types in Acrobat in a single process. You cannot add a non-Acrobat file to a binder unless you have the source application on your computer.

Follow these steps to create a binder file:

1. Choose From Multiple Files from the Create PDF task button menu to open the Create PDF File from Multiple Documents dialog box.

2. Click Browse to display an Open dialog box. Locate the first file you want to use in the binder and click Add. The Open dialog box closes, and the file is added to the Files to Combine pane in the dialog box.

3. Repeat until all desired files are selected (Figure 5.2). The documents may be in different allowable file formats, and you can select the same file more than once.

4. Click Add all open PDF documents if you have open documents in Acrobat and want to include them in the binder; click the Include recently combined files drop-down arrow and select previously constructed binder files.

5. To reorganize the files, click a file in the Files to Combine pane, and then click a directional button to reorder the list or delete a file.

6. Click OK to close the dialog box and create the binder. Each file is processed separately.

7. When the conversions are complete, name the file and choose its storage location in the Save As dialog box, and then click Save to save the combined document. The file is named binder1.pdf by default.

Note: You can preview PDF documents before building the file. Click the file's name in the Files to Combine pane of the dialog box, and click Preview to open a preview window. Use the arrow controls or type a page number to view pages in a multi-page PDF document; click OK to close the Preview dialog box.

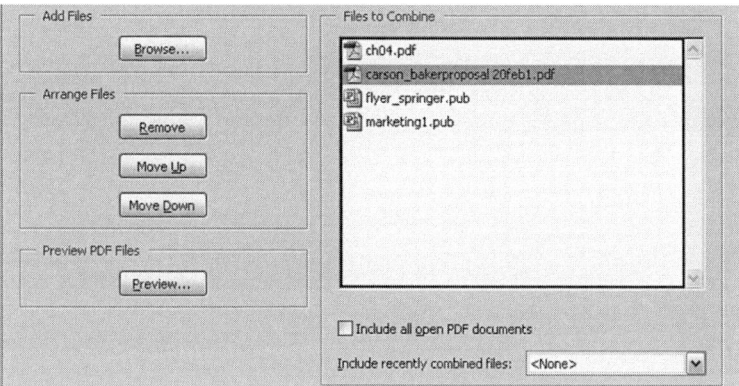

Figure 5.2 Combine several files to create a single PDF document

Converting a Web Page to PDF in Acrobat

Click the Create PDF task button to display its menu and choose From Web Page to open the Create PDF from Web Page dialog box (Figure 5.3).

Follow these steps to download and convert Web pages to PDF within Acrobat:

1. Define the file you want to convert in the dialog box by typing the URL for the file, clicking the drop-down arrow and selecting a previously opened Web page, or clicking Browse to open a dialog box to locate a file locally.

2. Specify the number of levels of the Web site you wish to download. The default is one level, which refers to the initial set of pages on a Web site. If you choose Get Entire Site, all pages of the site are downloaded regardless of the number of pages or number of levels.

3. Click Create to start the conversion. The Download Status dialog box displays and describes active connections, names, sizes, and locations of files.

4. The converted PDF file, complete with bookmarks, is displayed in Acrobat.

5. Save the file.

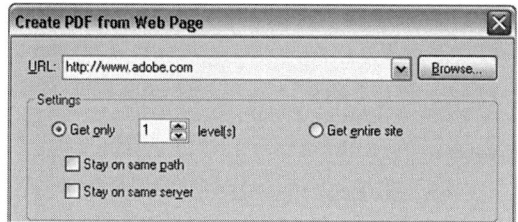

Figure 5.3 Specify conversion settings for Web pages

Headers and footers are added to the pages of the PDF document automatically. The header shows the Web page's name, while the footer shows the page's URL, number of pages, and download date and time (Figure 5.4).

Custom Web Page Conversions

Before converting Web pages to PDF, you can specify some page layout and other conversion settings. Click Settings on the Create PDF File from Web Page dialog box. The General tab of the Web Page Conversion Settings dialog box displays, showing the file types and PDF settings tab (Figure 5.5).

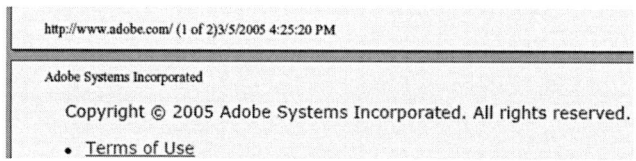

Figure 5.4 Acrobat applies headers and footers to converted Web pages by default.

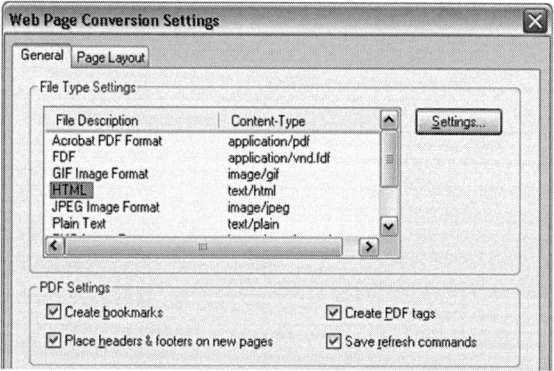

Figure 5.5 Select custom file type and PDF settings

Only HTML or Plain Text formats can be modified. Select the file type on the General tab and click Settings; choose features such as fonts and encoding, and click OK to close the Conversion Settings dialog box and return to the previous dialog box.

Click the Page Layout tab to specify common page layout options such as page size, margins, and orientation. You may want to modify page margins if you are combining a Web page PDF with other documents, for example. Click OK to close the Web Page Conversion Settings dialog box and return to the Create PDF from Web Page dialog box.

Appending Additional Pages

To add additional levels to a PDF file converted from a downloaded Web site, choose Advanced > Web Capture > Append Web Page. The Add to PDF File from Web Page dialog box opens, which is identical to the Create PDF File from Web Page dialog box shown in Figure 5.3. Select another level to add pages to those already in the document.

   You can also work directly from your existing converted Web page PDF file. Move the mouse over an active Web link and the cursor changes to a Web pointer ⬟; right-click and choose a command from the shortcut menu for converting Web pages associated with the link.

Creating a PDF from a Scan in Acrobat

If you have a printed copy of a file, you can still create a PDF version of it using Acrobat's scanning feature. Follow these steps to capture a scanned document in Acrobat:

1. Click the Create PDF task button to display the menu and click From Scanner to open the Create PDF File From Scanner dialog box (Figure 5.6).

2. Select a scanner from the Device drop-down list and choose single- or double-sided scanning. Select other options, such as recognizing the text as OCR, or specify that the scanned document is tagged.

3. Click Settings to open a dialog box for changing conversion options such as PDF output style and image downsampling. Make the modifications and click OK to close the dialog box.

4. Specify a destination for the scanned page, either a new document or an open file in Acrobat to which you can append the scanned document.

5. Click Image Settings to open a dialog box to specify image conversion options such as compression and filters. Make the modifications and click OK to close the dialog box.

6. Click Scan to start the conversion process; follow the prompts, which vary according to your scanner and its software.

7. The scanned document is opened in Acrobat; save the PDF file.

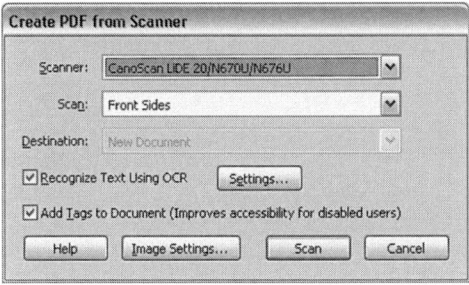

Figure 5.6 Scan a document directly into Acrobat as a PDF file

   By default, files scanned and converted to PDF create a searchable document containing both words and images. An image PDF displays the content of a page as a single image. Image PDFs are generated by programs such as Photoshop, as well as by older versions of Acrobat. To examine a file to see which type of PDF you are working with, choose the Select tool ⬚ on the Basic toolbar and then click some text on the file. If you see the vertical I-beam cursor, the page contains text; if the entire page is selected, your file is an image PDF.

Converting Scans Outside Acrobat

Many scanner companies have licensed the Adobe library or a third-party library to produce PDF scans directly from the scanner. In a big production job, it is often simpler to create PDFs through the scanner than through the Acrobat OCR process.

Producing Editable Text from an Image PDF

Acrobat uses optical character recognition (OCR) to convert an image to searchable text and images; you can choose several ways to convert the contents of the file. Searchable image is by far the most commonly used OCR method in AEC work. The result is an exact, but deskewed copy that is also searchable. For legal purposes, it is the only format to use.

It is not necessary to convert all image files to searchable text and images. Leave the PDF file as an image unless you intend to make the text available to assistive devices such as screen readers, manipulate the content (such as revising text, extracting tables, or export in different formats), or need to search the text. Follow these steps to capture the content of an image PDF:

1. Choose Document > Recognize Text Using OCR > Start to open the Recognize Text dialog box. Specify the page range to which the OCR capture is applied.

2. Click the Edit button to open the Recognize Text – Settings dialog box (Figure 5.7). Choose output and downsample image options and then click OK to return to the Recognize Text dialog.

   The choices include:

   - **Searchable Image.** This option keeps the foreground of the page intact and places the searchable text behind the image.
   - **Formatted Text & Graphics.** This format rebuilds the entire page; the content is converted to text, fonts, and graphics.
   - Click the **Downsample Image** drop-down arrow and choose a downsample size ranging from 600 to 72 dpi.

3. Click OK to start the capture process. The processing time varies depending on the complexity and size of the document. The dialog box closes when the document is captured. Once the capture is complete, evaluate the file to assess the accuracy of the capture.

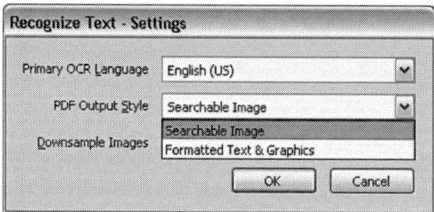

Figure 5.7 Choose conversion options for the output style and image downsampling

Tips for Easier Document Capture

Sometimes you have no choice with regard to the quality of a document that is being scanned and converted. For simpler post-capture correction in your document keep these tips in mind:
- Acrobat won't scan a document for OCR unless the resolution is set at a minimum of 144 dpi.
- Scan for black and white at 200 – 600 dpi and scan at 200 – 400 dpi for grayscale or color.

- Use OCR fonts or other clear fonts at about 12 points if possible; black text on a white background is the best for accuracy, while colored backgrounds and decorative fonts cause the most errors.

Evaluating Suspects

If you chose formatted text and graphics as the OCR method, once the capture is complete you can evaluate the results of the conversion. Characters that cannot be definitively converted are defined as *suspects*. To view the suspects in a document, choose Document > Recognize Text Using OCR > Find All OCR Suspects. Any content that needs confirmation is outlined with red boxes.

Choose Document > Recognize Text Using OCR > Find First OCR Suspect or select the TouchUp Text tool ⬚ on the Advanced Editing toolbar and click a suspect on the document to open the Find Element dialog box (Figure 5.8).

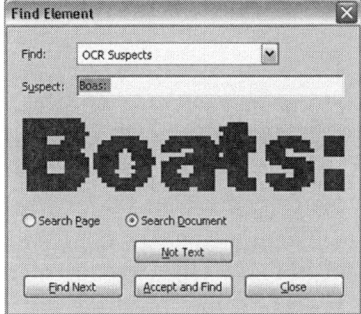

Figure 5.8 Evaluate suspect items in this dialog box

In the figure, the word "Boats" is suspect, and the OCR process identified the text as "Boas:" Type replacement text in the Suspect field to make changes; click Accept and Find to confirm the interpretation and go to the next suspect. Click Not Text if the object is not text; click Find Next to view the next suspect; and click Close to end the evaluation process.

You can use the TouchUp Text tool to modify fonts that may have changed if you used the Formatted Text & Graphics conversion option. Save the document.

Creating a PDF from a Clipboard Image

You can copy and paste images to your system's clipboard to use to create a new PDF document. In Acrobat, you can either create a new file or add clipboard contents to an existing file as a stamp.

Creating a New PDF File

Select and copy the image you want to use for a new PDF document in its source program. In Acrobat, you can choose Window > Clipboard Viewer to see the copied content. To produce a new PDF file, choose From Clipboard Image from the Create PDF task button's menu. The clipboard content is processed and the new file opens in Acrobat.

If your source file was a JPEG image, the Picture Tasks plug-in and its associated information dialog box may display – click OK to dismiss the dialog box.

Pasting a Clipboard Image as a Stamp

Rather than creating a new PDF file, you can use an image copied to the clipboard as a stamp in an existing document. The pasted image is a type of comment. Follow these steps to paste the image as a stamp:

1. Select and copy the image from your source program or from Acrobat using the Snapshot tool 🔲 on the Basic toolbar.

2. Open the PDF document in which you want to paste the content.

3. Choose Tools > Commenting > Paste Clipboard Image. The mouse cursor changes to a Stamp 🖏 cursor.

4. Click the page where you want to insert the image; the position you click on the page defines the center of the pasted image.

Once the image is placed on the page, you can reposition it by dragging on the page, or resize it by dragging a resize handle.

Converting Web Pages to PDF

In addition to converting a Web page from within Acrobat, you can also create a PDF file directly via the PDFMaker installed by Acrobat 7 into Internet Explorer. The commands are also accessible by right-clicking the page to display the shortcut menu.

Click the drop-down arrow on the Adobe PDF toolbar and choose Convert Web Page to PDF from the menu (Figure 5.9). In a framed Web page, all content is flattened into one PDF page.

Rather than creating a new document, you can attach a Web page to an existing PDF file by following these steps:

1. Click Add Web Page to Existing PDF from the Adobe PDF toolbar to display the Open dialog box.

2. Locate and select the document to which you want to attach the Web page.

3. Click Save to convert the Web page and append it to the end of the selected document.

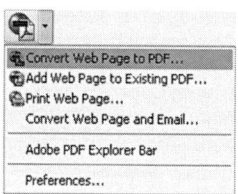

Figure 5.9 Internet Explorer includes a PDFMaker

The Internet Explorer PDFMaker includes an interesting feature for locating files. Click the Adobe PDF toolbar and select Adobe PDF Explorer Bar. A pane opens at the left of the Web browser window (Figure 5.10). PDF documents opened from the Adobe PDF Explorer Bar display in Adobe Reader within Internet Explorer. Close the Explorer Bar when you are finished working with it, as it will display the next time you open the browser.

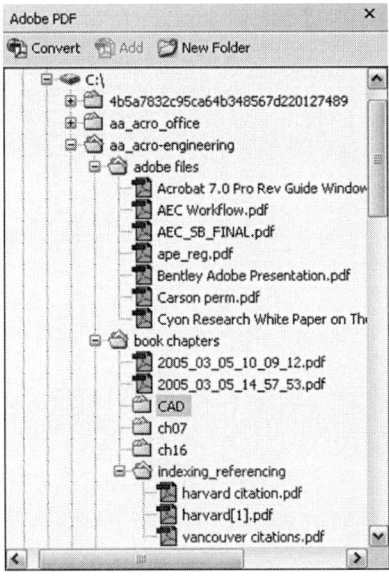

Figure 5.10 Search for files in a browser using the Adobe PDF Explorer Bar

Modifying Acrobat's PDF Creation Settings

Rather than opening a source program and choosing conversion settings, you can modify many settings for conversion directly through Acrobat's preferences.

Follow these steps to modify the settings:

1. Choose Edit > Preferences to open the Preferences dialog box. Click Convert to PDF in the column at the left to display the options in the right pane of the dialog box.

2. Click a file format in the Converting to PDF listing to display its settings; Figure 5.11 shows the conversion settings for a TIFF image. Some file formats, such as GIF, HTML, JDF Job Definition File, JPEG2000, and TXT, have no customizable options. In those cases, the Edit Settings button is disabled.

3. Click Edit Settings to open an Adobe PDF Settings dialog box. The content of the dialog box varies according to the active file format. Image file types may display options for Compression and/or Color Management, as shown in Figure 5.12.

4. Configure the settings as required, and then click OK to close the Edits Settings dialog box; click OK to close the Preferences dialog box.

Other conversion settings, such as those for PostScript/EPS files or Microsoft Office files, show the Adobe PDF Settings for supported documents dialog box when the Edit Settings button is clicked (Figure 5.13). Choose conversion settings and common options such as accessibility and reflow. Click Edit to open the PDF Settings dialog box to choose custom options; customizing PDF Settings is described in Chapter 4.

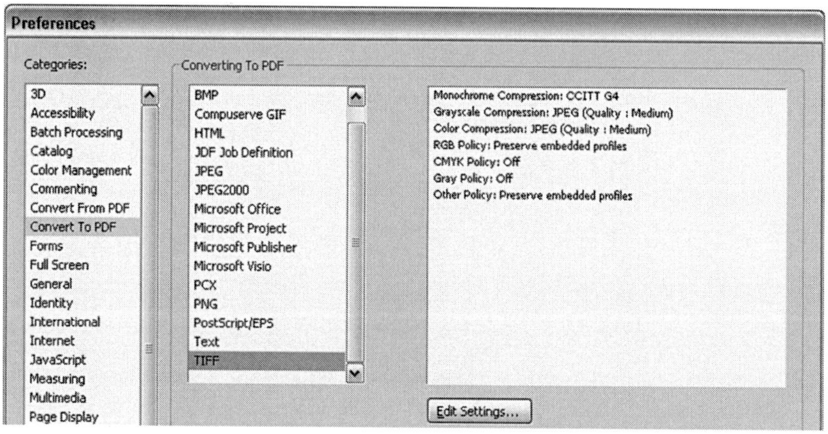

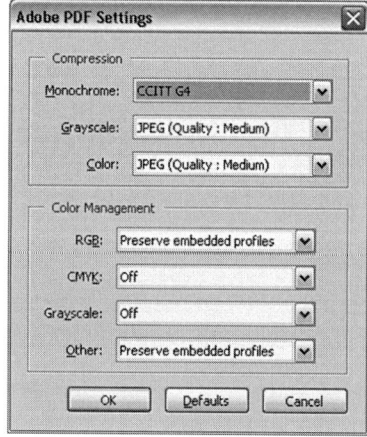

Figure 5.11 Configure file conversion options in Acrobat's preferences

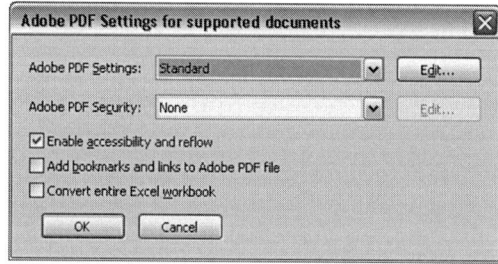

Figure 5.12 Choose custom image conversion settings

Figure 5.13 Select conversion options and security settings from this dialog box.

Summary

In this chapter we looked at several methods used within Acrobat to create PDF documents without having to open up source programs. You saw how Acrobat is used to convert single files, multiple files, scans, Web pages, and captured content.

Scanned content is captured using OCR, and then Acrobat can be used to evaluate and assign values to suspect characters. You also learned how to create a PDF file from a Web page using Internet Explorer's PDFMaker. Finally, ways to modify conversion settings using Acrobat's preferences were identified.

Exercises

1. Using the methods outlined in the chapter, convert various single and binder PDF files.
2. Compare and contrast the processes used to create a PDF file from a Web page within Acrobat and within Internet Explorer using its PDFMaker.
3. Experiment with scanning and capturing text in a PDF file in Acrobat.
4. Using different resolutions when scanning, evaluate the efficiency of the capture process.

Copyright Notice: Many elements of this project have been modified due to copyright and liability issues, and should not be used in other projects. They are only being used to teach the processes in the context of this book.

Project

In Chapter 4 we created a PDF cover page for a project specification. The goal in this project is to create more components of the project. Use the source files in the **ch05_project** folder.

Task 1: Creating PDF files from within Acrobat

1. Open the **0_Cover.pdf** file created in Chapter 4; a copy of the PDF file is also included in the **ch05_project** folder.
2. Click the Create PDF task button and choose From Multiple Files to open the dialog box.
3. Click the Include all Open PDF documents checkbox; the **0_Cover.pdf** file is listed in the Files to Combine Box.
4. Click Browse. Locate your **ch05_project** folder and select:
    *   corps-tva permit application blank.PDF
    *   03300 Cast-In-Place Concrete.DOC
    *   03305 Concrete Curing and Finishing.DOC
    *   04220 Concrete Unit Masonary.DOC
    *   Bid Tabulation Sheet.XLS
    *   Photo of Field.JPG
5. Click the **Photo of Field.JPG** file in the list, and click Move Up to move the image before the XLS file in the list.

6. Click OK. Acrobat generates the combination file, and assigns the name **Binder1** by default. Change the name to **Spec Book**, and save the file.

Task 2: Adding form content

1. Click the Bookmarks tab to show the Bookmarks pane; you see each document has a corresponding bookmark.
2. Click the **corps-tva permit application blank.PDF** bookmark to display the page.
3. The application is an active form. Click the fields and add content to the form.
4. Save the file.

Task 3: Creating PDF files from scans

1. In the **ch05_project** folder, locate and open the **Bid Sheet to Scan.PDF** file. Print it to paper and close the file.
2. In Acrobat, open the **Spec Book.PDF** file.
3. Place the printed copy of the **Bid Sheet to Scan.PDF** in your flatbed scanner upside down and skewed (as much as 15 degrees.)
4. Cick the Create PDF task button to open its menu and choose From Scanner to open the dialog box.
5. In the Create PDF from Scanner dialog box choose the Append to Document destination, and make sure Recognize Test using OCR is checked.
6. Click Settings; in the Recognize Text Settings dialog box choose Searchable Image from the PDF Output Style drop-down list. Click OK.
7. Click Scan to start the scanning process. In your scanner's dialog box, choose Black and White and 300dpi options.
8. When the page's scan is complete, choose Finished. Acrobat makes the text searchable, rotates the page to the correct view and corrects the skew automatically.
9. Click the Pages tab to display the Pages pane. You see the page is appended to the end of the document.

Task 4: Creating PDF files from web pages

1. Open the **Spec Book.PDF** file in Acrobat.
2. Click the Create PDF task button to open its menu and choose From Web Page.
3. In the URL field, type http://www.donnabaker.ca/ and leave the default 1 layer setting in the dialog box.
4. Click Create. Acrobat converts the HTML to a separate PDF named using the Web pages' names and maintains all links.

Task 5: Adding the Web pages

1. With both the **Spec Book.pdf** and converted Web page PDF files open in Acrobat, choose Window > Tile> Vertically to display both files.
2. Click the Pages tab in each file to show the Pages panes. Press Ctrl+click all three thumbnails in the **Donna Baker** file to select them.
3. Drag the thumbnails to the end of the **Spec Book.PDF** file's thumbnails.
4. Save the **Spec Book.PDF** file. The file saved to this point is used as the source file in Chapter 7's project.

Task 6: Creating a PDF file from a clipboard image

1. Click the Bookmarks tab to open the Bookmarks pane in the **Spec Book.PDF** file.
2. Click the bookmark for the Photo of Field.JPG to display the page in the Document pane.
3. Use the Zoom tool to magnify any part of the picture several times. Do not set the magnification higher than approx. 200% as the image pixelates at that level.
4. Click the Snapshot tool on the Basic toolbar to select it.
5. Drag a marquee around a portion of the image; release the mouse to capture the content and place it on the clipboard.
6. Click the Create PDF task button to open its menu and choose From Clipboard Image to convert the captured content to a separate PDF file.
7. Do not save the file; it is not required in the project.

6

Converting AEC File Formats

One critical factor in an AEC project's success is the ability to communicate efficiently with all stakeholders involved in the project. As in many areas of modern business life, converting content to PDF format is a fairly simple process. In other areas, such as the AEC world, in addition to conveying information written, verbally, and numerically, one of the most important means of communication is by use of technical drawings.

Although there are certainly other programs on the market, in this book we concentrate on three programs. This chapter describes how to create PDF files from AutoCAD, and Bentley MicroStation files; Microsoft Visio is included in Chapter 4. In addition to using the drawings themselves, Acrobat also allows you to convert and use the metadata content attached to the components of a drawing.

We are very pleased to offer content and projects written by noted authorities on Acrobat and PDF, and are grateful for their participation. This chapter looks at converting files to PDF from AutoCAD. The information is provided by Tim Huff, Acrobat's Business Development Manager in the AEC space.

We include information on producing PDF files in Bentley's MicroStation, as well as viewing and using 3D PDF files, contributed by Jo Terri Wright, the Publishing Specialist at Bentley.

In this Chapter

This chapter describes how to create PDF files in common AEC applications, the raison d'etre of this book. You will learn how to:

- Use the AutoCAD PDFMaker
- Use Bentley products to convert design files to PDF
- View and use U3D files
- Understand and manipulate layered files in Acrobat.

Creating Round-trip PDFs – Tim Huff, Adobe Systems

Creating round-trip PDFs from AutoCAD and Acrobat are as easy as 1-2-3, but what do I mean by "round-tripping" ?

The term describes exporting an AutoCAD DWG file as a PDF file, sending the PDF file out for review, and bringing those comments back into the DWG file within AutoCAD. The round-trip process works with AutoCAD 2002, 2004, 2005, and 2006. Users can participate in the review using Acrobat Professional 7, Acrobat Standard 7, or if allowed by the initiator, Adobe Reader 7.

There are substantial benefits to using a round-trip method:

- Participants do not need AutoCAD to be part of the review process, which allows field personnel and "non-CAD" users to add comments on the design.
- It allows for a true review process that is easily done with any MAPI-compliant email system (for this chapter we will assume Microsoft Outlook).
- It allows the comments to be brought back into AutoCAD for ease of editing.

Practice File

This demonstration uses a file named **WestSidePoliceStationACAD 2000.dwg** available from the book's Web site. The drawing used in this demonstration is © 2004, John TeSelle of JT Architects.

Publishing the Drawing

We will start with an AutoCAD Drawing. Figure 6.1 shows a drawing of a police station, but it could easily be any DWG file, of course.

The steps we are going to take follow the standard procedure for the conversion process. We will publish the PDF file from AutoCAD using the PDFMaker buttons.

The three buttons represent three ways of creating the PDF from AutoCAD:

- **Convert to Adobe PDF** . This method creates a PDF file from an AutoCAD DWG and saves it locally to your hard drive
- **Convert to Adobe PDF and EMail** . This method creates a PDF from the AutoCAD DWG and saves it locally, then invokes your email client, places a standard notation about the PDF in a new email message, and then emails the PDF to the selected recipients
- **Convert to Adobe PDF and Send for Review** . Use this method to create a PDF file from the AutoCAD DWG file and save it locally. Then your email client is invoked, and dialog boxes display for you to set up a review cycle by adding users to the list. The command also places a standard notation about the review in the email message.

Follow these steps to convert the file and start the review:

1. Open AutoCAD and the DWG file.
2. Click the Convert to Adobe PDF and Send for Review button on the PDFMaker toolbar (Figure 6.2).
3. The Acrobat PDFMaker wizard opens and lists the layouts defined in the AutoCAD drawing. You can add both Model and Paper space views to your PDF file (Figure 6.3).

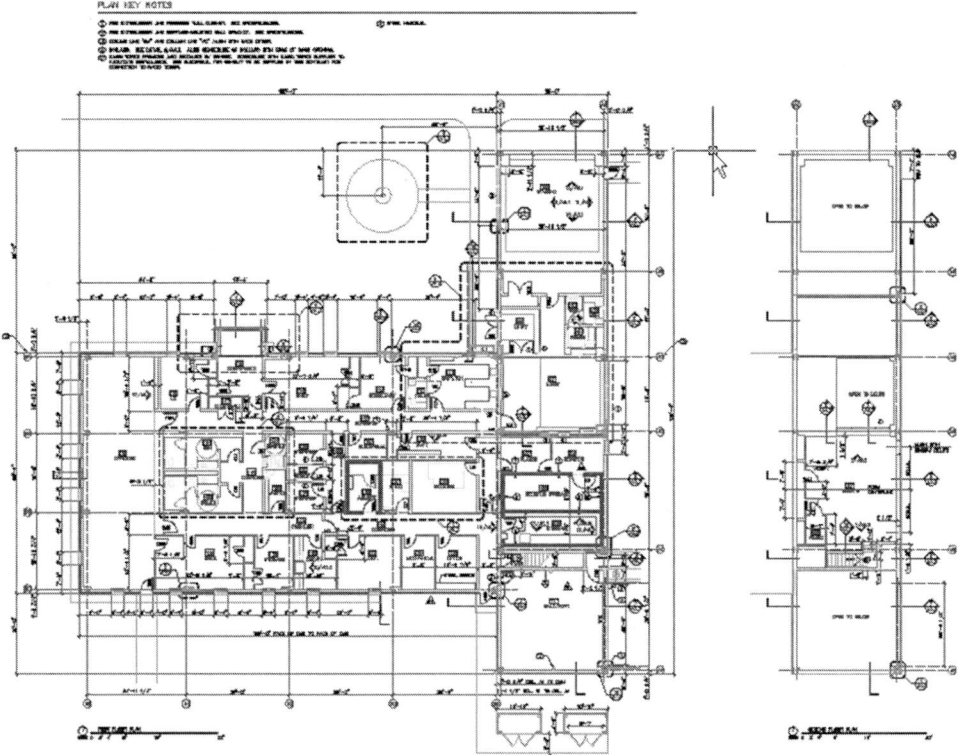

Figure 6.1 Police station drawing to be converted to PDF

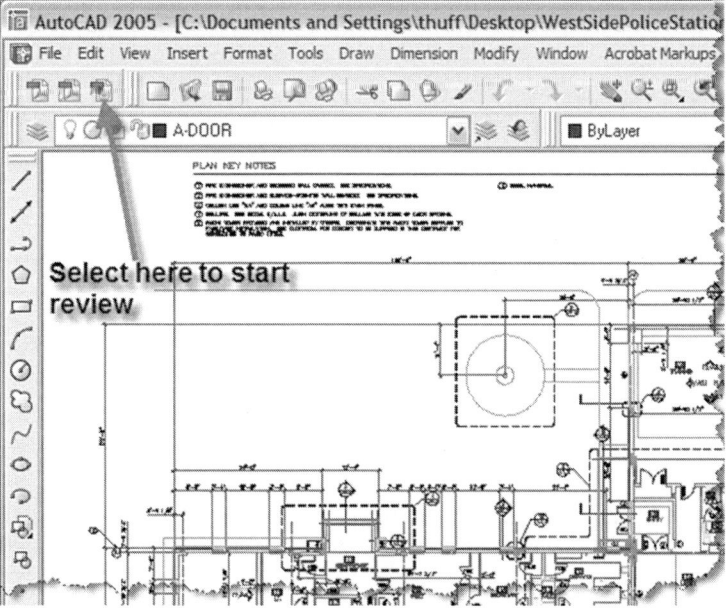

Figure 6.2 Use the PDFMaker tools

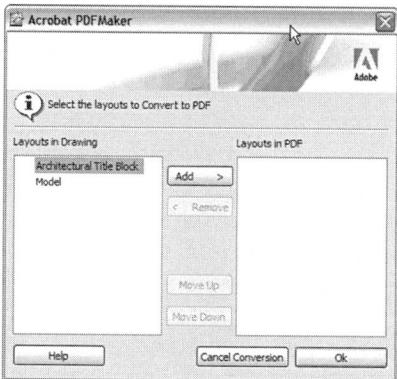

Figure 6.3 The wizard lists the views

4. Click a view in the Layouts in Drawing column then click Add to move it to the Layouts in PDF column. In our example, we add both Model space (named *Model*) and Paper space (named *Architectural Title Block*).

5. Reorder the Layouts in PDF list if desired; click a layout and then click Move Up or Move Down. In our example, Model is moved to the top of the list.

6. Click OK to display the next pane of the wizard. On this pane, define how the layer structure will be exported from the DWG file. In the example, I selected Retain all or some layers, which allows me to use the rich data inside the DWG file (Figure 6.4). To strip the layer structure from the PDF file, select Flatten all layers.

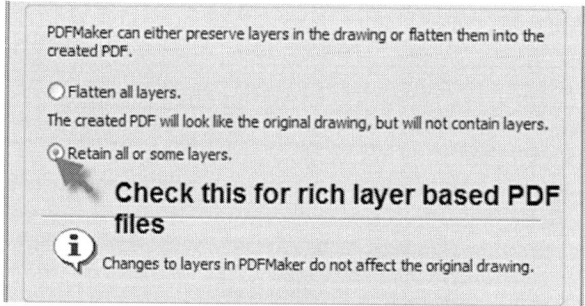

Figure 6.4 Select the layer structure for the PDF

7. Click Continue to display the next pane of the wizard, where we will add more intelligence to the PDF file. In Figure 6.5, you see all the AutoCAD layers in the left column.

8. Right-click the column at the left of the wizard to open a shortcut menu and choose Select All; then click Add Layers to copy the list to the right column (Figure 6.6).

9. In this case, our AutoCAD DWG file did not include any Layer Groups, but we can easily add them in the wizard. Shift-click to select the layers for grouping in the right column on the wizard, and then right click and choose Create a new Layer Set from the shortcut menu. Type a name for the layer set: ours is named Walls.

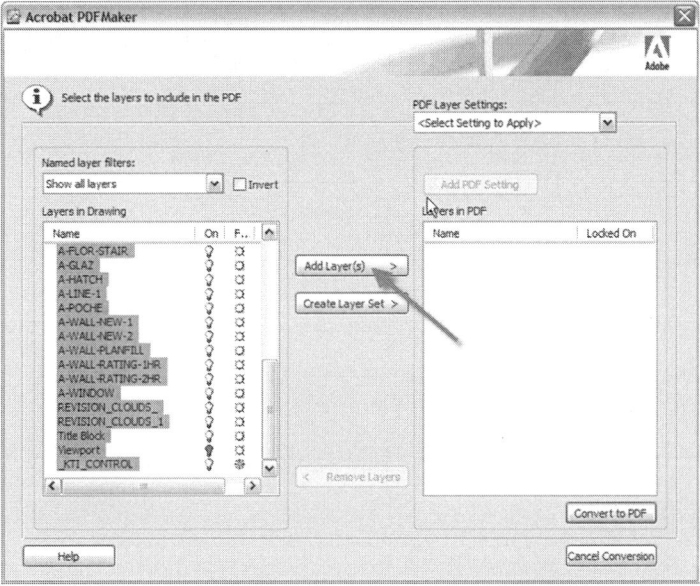

Figure 6.5 Layers are listed in the wizard

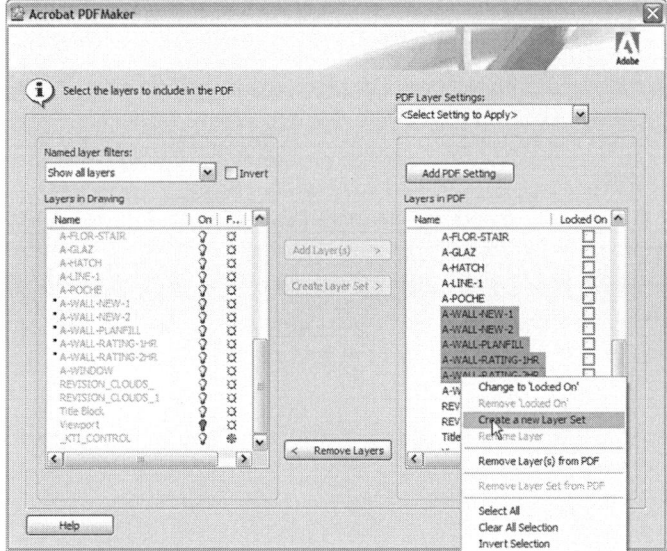

Figure 6.6 Select layers and group as a layer set

10. Once you have created the layer group, you can save it for later use. Click Add PDF Setting; in the small dialog box that opens, name the setting. Ours is named DWG LAYOUT; click OK to close the dialog box and save the setting (Figure 6.7).

11. Finally, click Convert to PDF. A Save As dialog box opens. The PDFMaker uses the AutoCAD file name and location by default – change the name and storage folder if desired and click Save.

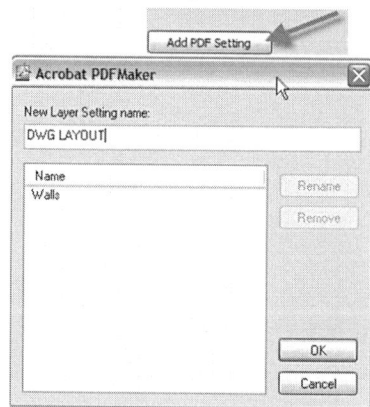

Figure 6.7 Save PDF settings for future use

Note: The Send by Email for Review wizard is described in depth in Chapter 12.

Issuing Review Invitations

Once the file is saved, the Send by Email for Review wizard displays, since we used the Convert to Adobe PDF and Send for Review option on the PDFMaker. The wizard guides you through the creation of an email-based review cycle.

You can add as many people to the review as needed, and if you are using Acrobat 7 Professional you can open up the review to someone that only has the free Adobe Reader 7! This increased use of the reviewing tools allows for non-CAD and non-technical reviewers to participate in the review cycle.

Follow through the wizard:

1. In Step 1, select the PDF file that is the subject of the review. In our example, the newly converted AutoCAD file is already selected.

2. In Step 2 we get to the heart of building the intelligent document work flow and add reviewers to our Email list. If you are using a MAPI-based email system, click the Address Book button to display your Address books (Figure 6.8). I am very confident in my work, so I am going to review my own work here!

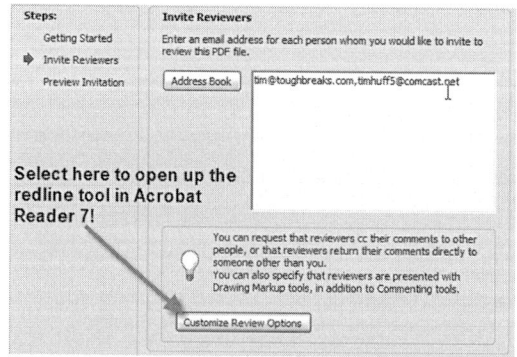

Figure 6.8 Add addresses to invite reviewers

3. Also in Step 2 of the wizard, click the Customize Review Options button to open a dialog to specify additional usability aspects of this work flow. I can add more intelligence to my PDF file by automatically displaying the Drawing Markup Tools. Most importantly, I can enable Drawing Markup inside the free Adobe Reader 7 program, which can allow people outside the firewall or without a copy of Acrobat 7 Professional to participate in the review (Figure 6.9).

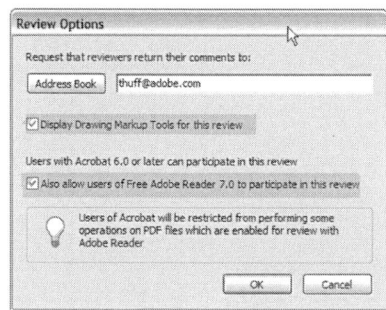

Figure 6.9 Specify additional rights

4. Click OK to close the Customize Review Options dialog box, and then click Next to move to the third pane of the wizard.

5. Review the automatically generated email that will be sent to your reviewers, and edit as necessary. The email also supplies a link to the Adobe Reader download in the event the reviewer does not have a current version of the software.

Receiving the Invitation

Now let me switch roles to one of the reviewers. In Figure 6.10 you can see that my inbox shows an email inviting me to a review cycle. I simply open the email and double click the attachment to open the PDF. In this case, I only have Adobe Reader 7.

As you see in Figure 6.11, the Adobe Reader program shows a number of items that do not exist in a standard PDF file:

- The Document Message Bar across the top of the graphics area gives hints and lets you know why you are seeing all these new user tools.
- The "How to…" shown at the right of the window walks you step-by-step through the tools and explains how to participate in the email review.
- You see the Drawings Markups tool bar is shown at the top of the screen.
- Notice on the left of the screen that all the rich layer data from the AutoCAD file are present. I can turn layers on and off as needed for my review!

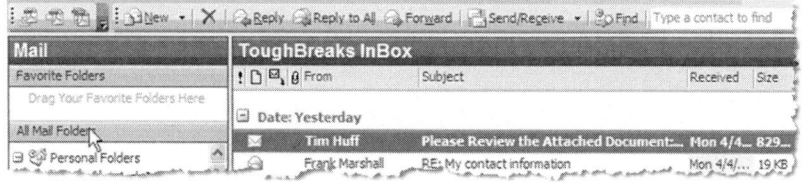

Figure 6.10 The invitation arrives in the recipient's inbox

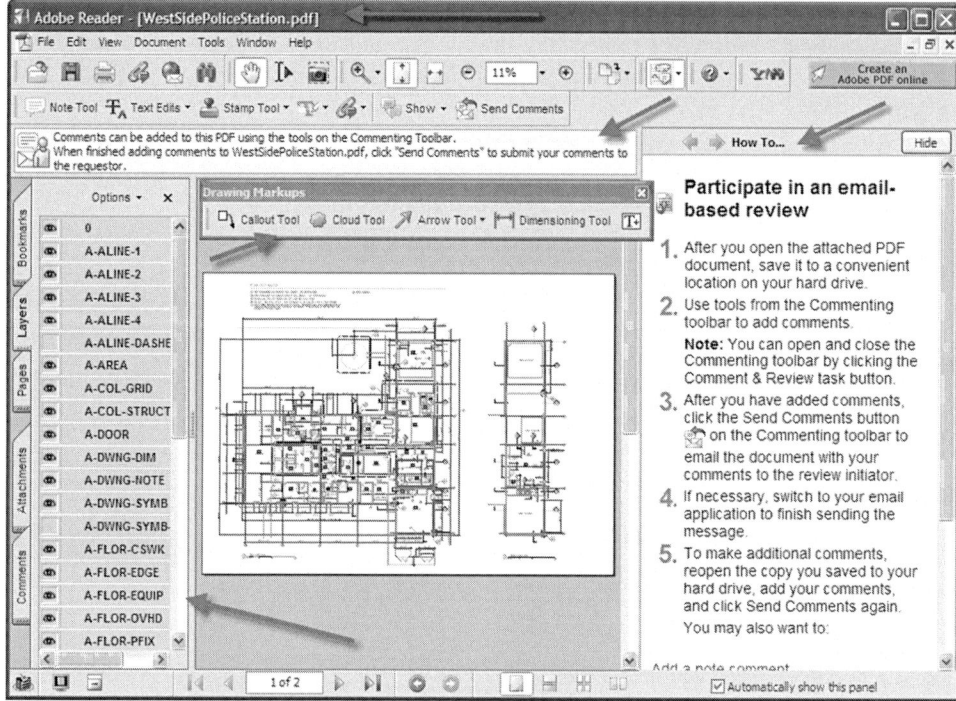

Figure 6.11 A drawing enabled for use in Adobe Reader

Markups and Returns

Acrobat has always offered markup tools, but the AEC and Mechanical Engineering standards are different and use different ways of marking up drawings using lines, arcs, clouds, and notes to identify the edits to the originator. Acrobat 7 Professional includes a suite of tools that let the reviewer convey his/her edits in the standard fashion. Figure 6.12 shows the start of my review. Click the Arrow tool's drop-down arrow to open the Drawing tools' menu, which reveals a wealth of tools for adding drawing-based annotations.

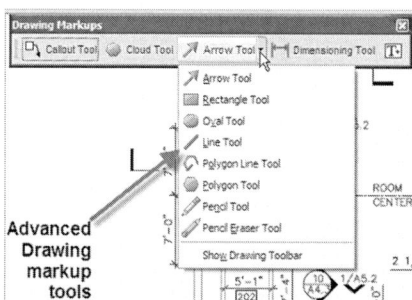

Figure 6.12 Drawing tools provide standard markup options

For example, I used the Arrow tool to create arrows with annotations that point to my area of interest, then used the Pencil tool to sketch the door's approximate location, finally using the Cloud tool to draw attention to the area I edited (Figure 6.13).

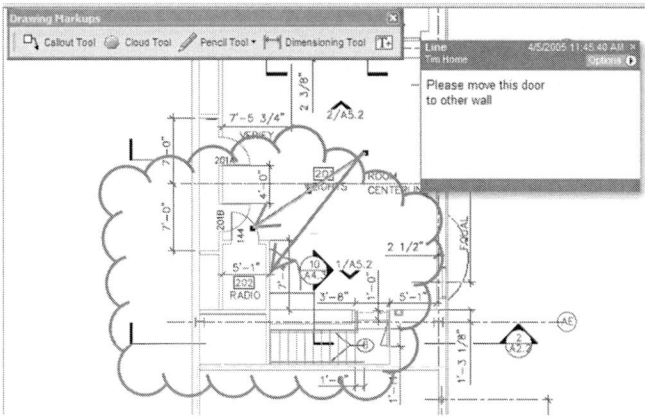

Figure 6.13 Use annotations to make edits

Now that I am done with my review, I need to send my comments back to the originator. I can either click the Send Comments button on the Commenting toolbar or choose File > Send Comments. The command automatically opens an email dialog box that includes a preformatted message and the return address for the originator. The message can be edited as needed.

Integrating Annotations

I now return to the role of originator. You can see in Figure 6.14 that I received two emails with comments. I simply open the email and double-click the PDF attachment.

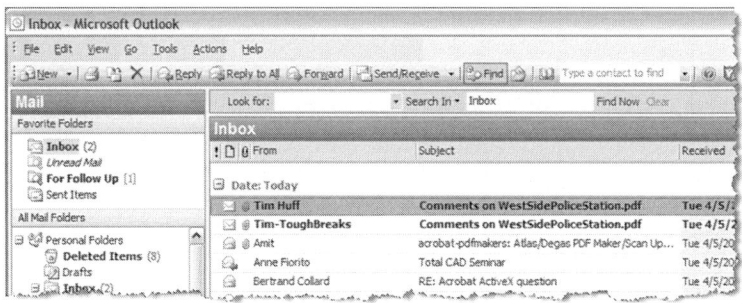

Figure 6.14 Open comments directly from an email

Since I originated the review, Acrobat displays the Merge Comments dialog box, asking if I want to bring the comments back into my original PDF file or open the copy sent to me. I select Yes to import comments from both emails.

The markups are integrated into my original file (Figure 6.15). Although I can manipulate the comments in many ways, I cannot edit the PDF file. At this point, the round-trip comes into play.

Editing the PDF File

If I open the original CAD file in AutoCAD, one of the Acrobat menus installed with Acrobat 7 Professional is the Acrobat Markups menu. Choose Acrobat Markups > Import Comments from Acrobat. In the Import Comments from Acrobat dialog box, select the comment files to integrate, and click Continue (Figure 6.16). The comments from the PDF file are actually integrated into the original drawing on a new layer with new comments objects!

> **Note:** The comments are placed on their own layer in AutoCAD and can be hidden after edits have been made, thus allowing for revision traceability. This fulfills ISO standards, and most PDM products.

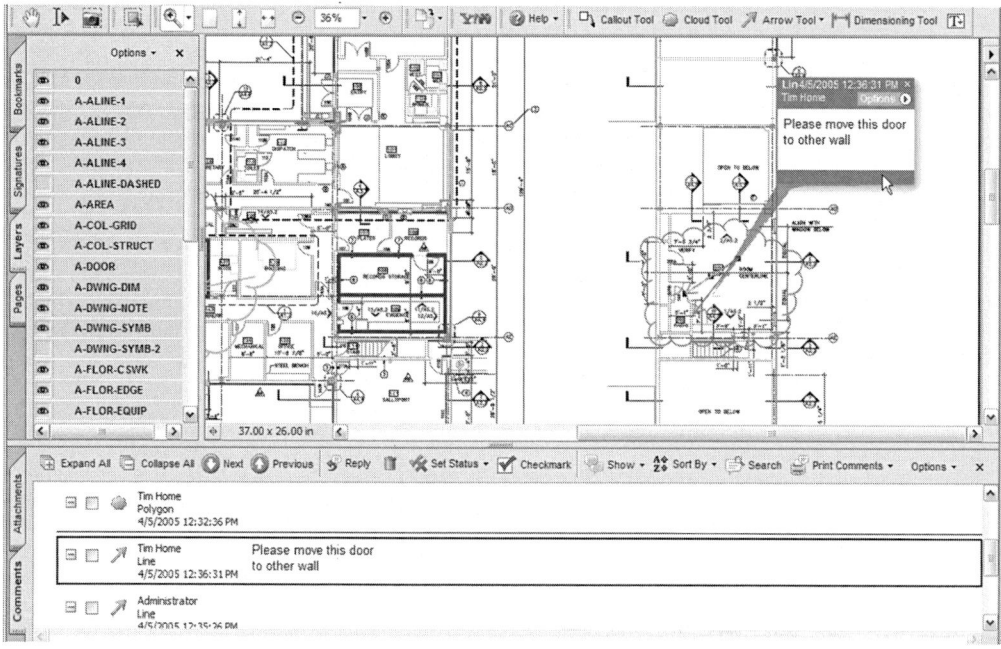

Figure 6.15 Markups are integrated into the original file

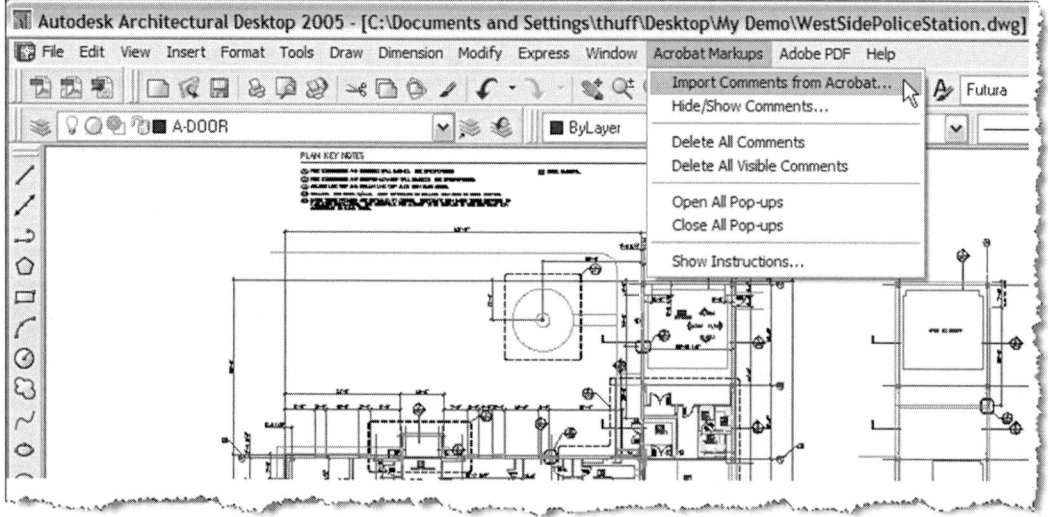

Figure 6.16 Import comments into the original drawing

In this demonstration, we took an AutoCAD drawing and, using Acrobat Professional 7, created a review workflow that did not require the reviewers to have AutoCAD – or Acrobat Professional 7 for that matter.

Extending the review to those working with Adobe Reader 7 opens up the review cycle for facilities, owners/operators, construction personnel – virtually anyone! Good luck with your new simple and easy review cycles!

MicroStation to PDF – Jo Terri Wright, Bentley Systems

Bentley Systems, Incorporated partnered with Adobe Systems to publish authentic Adobe PDF documents directly from their MicroStation and ProjectWise product lines. Bentley publishes PDF documents from DGN and DWG files which can contain vector and raster data as well as 3D models. All that is needed in order to view any PDF created with the Bentley products is Adobe Reader.

MicroStation users produce authentic PDF documents using the print and batch print interfaces the same way they print to paper or other formats so there is nothing new to learn. These PDF documents can retain file and level information, include engineering links, searchable text, 3D models, a password to open the document, and bookmarks for multi-page documents.

The intelligence that can be added to the PDF document increases when it is created using MicroStation PDF Composer, free to all MicroStation V8 2004 Edition SELECT users. MicroStation PDF Composer includes all the same features as MicroStation's PDF functionality plus the ability to add digital signature fields, specification documents, user defined intra-document links, and bookmarks for external links, points of interest, and embedded documents.

Using Bentley's AEC managed environment ProjectWise, users can include links in the PDF document that point to the originating design file or components within the design that are stored in ProjectWise. This enables authorized users the ability to see the current state of a project. With ProjectWise, generation of Adobe PDF documents can be automated to run at specific times, upon workflow-triggered events, or triggered by a design change.

PDF Creation from MicroStation

MicroStation printing offers users the ability to print to a Windows printer or to a collection of printers delivered with the software. Bentley now delivers a printer for generating Adobe PDF documents called pdf.plt. This printer enables users to take advantage of their printing setup and existing pen tables to create PDF documents. PDF options are given in the PLT file for turning on or off the ability to retain file and level information, engineering links, searchable text, bookmarks for multi-sheet PDF files, and for setting a password to open the document.

The PLT file can be copied several times and modified to eliminate the need to modify the printer each time you need different options. For example, PDF_standard.plt has all options turned off, PDF_level_search.plt has levels and searchable text on, and PDF_all has all options on.

Practice File

This demonstration uses a file named **BikeFrame.dgn** available from the book's Web site. Download the file to follow along with the discussion.

Selecting and Configuring the PDF Printer

Once you select and configure a PDF printer, you won't have to go through the configuration again unless you need to change a PDF output option.

Follow these steps to choose and configure the desired PDF printer:

1. In MicroStation V8 2004 Edition open BikeFrame.dgn.

2. Choose File > Print to open the Print dialog.

3. Choose File > Select Bentley Driver from the Print dialog and choose the pdf.plt printer (Figure 6.17).

4. Choose File > Edit Printer Driver from the Print dialog to edit PDF.plt and turn on searchable text and any other feature you would like on.

5. Save and Exit Notepad.

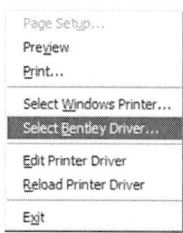

Figure 6.17 MicroStation print dialog file pull-down

Publishing the PDF from the Print Dialog

Unless you change your printer, each time you open the print dialog the PDF printer will be selected. If you're printing a sheet model, paper size, print scale, and so on will be set correctly in the dialog.

Follow these steps to print the PDF document from the Print dialog box:

1. Click the General Settings, Area pull-down and choose Fit All so that all the elements in the file print (Figure 6.18).

2. Click the Paper Size pull-down and choose ANSI E.

3. To maximize the print, either click the Maximize icon or choose Settings > Maximize.

4. Click the Auto-center check box for the Print Position.

5. Choose File > Print from the Print dialog.

6. On the Save Print As dialog box, select a folder to write the PDF document, key in a file name and select OK.

Practice Files

This demonstration uses files named **ar1.dgn**, **ar2.dgn**, and **learn.dgn** and a Design Script named **Composer.pen** available from the book's Web site. Download the files to follow along with the discussion.

PDF Navigation

Once the drawing is published as a PDF file, you can work with it in Acrobat or Adobe Reader 7. Here are some activities you can try with the drawing:

- Open the Layers tab and note that all the files and levels that were included in the PDF can be turned off and on – toggle several layers off and on.
- Search for the word Frame and note that two instances will appear in the results.

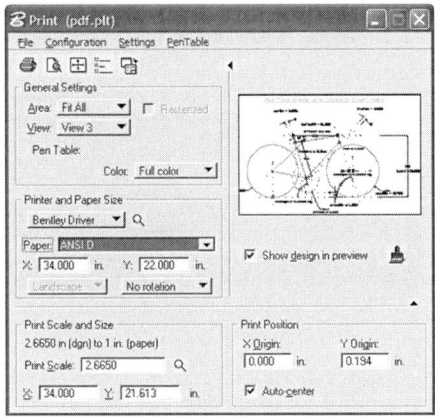

Figure 6.18 Set print options

PDF Creation from MicroStation PDF Composer

The MicroStation PDF Composer is available to MicroStation V8 2004 Edition SELECT users. Design files can have raster attachments and 3D models which can be represented in the PDF as interactive Universal 3D (U3D) models. MicroStation PDF Composer also supports DWG files.

Creating the Composer Set

PDF Composer is an interface that runs outside the CAD environment to publish PDF documents from both DGN and DWG files in a batch mode (Figure 6.19).

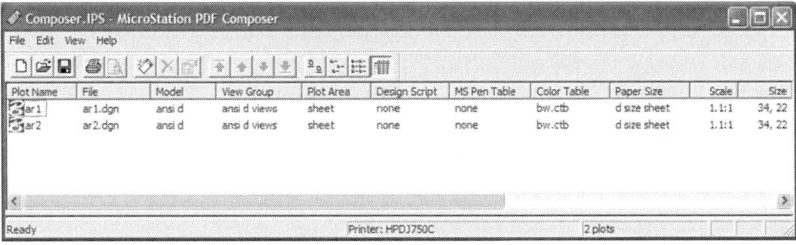

Figure 6.19 MicroStation PDF Composer Interface

Follow these steps to create the Composer set:

1. Open MicroStation PDF Composer and on the Welcome to MicroStation PDF Composer dialog select Create a new plot set from files you select and click OK.

2. Add the files **ar1.dgn**, **ar2.dgn**, and **learn.dgn** as input files on the Create Plots dialog and click OK. These files can be reordered in the list by selecting the appropriate files and clicking the move up and down arrow buttons.

3. Available form sizes are determined by selecting a printer on the Form Setup dialog. Choose File > Form Setup and select a wide-format printer such as an HP DesignJet. If one is not available, choose an available printer that has an 11 × 17 form such as a LaserJet.

4. Select all the DGN plots and click the Plot Properties button to display the Modify Plots dialog (Figure 6.20).

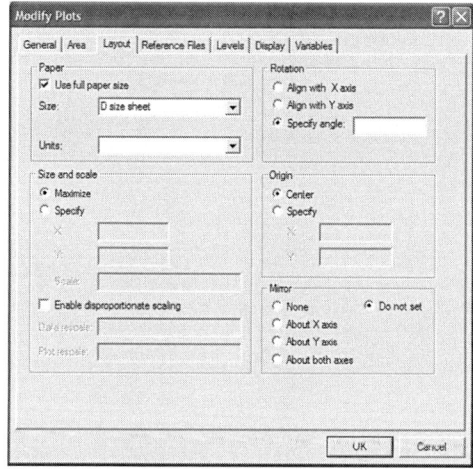

Figure 6.20 MicroStation PDF Composer Modify Plots dialog

5. Select a D-size form on the Layout Tab if available, if not select an 11 × 17 form and select Maximize to fit the plot to the form.

6. Select the General Tab on the Modify Plots dialog and attach the Design Script Composer.pen. This Design Script will take action on elements in the DGN files to add intelligence to the PDF document.

7. Select OK to save the settings.

8. Choose File > Save As to save the PDF Composer set. By doing this you save the sheets and their settings so that you can reproduce the PDF document and can even add sheets to the set as needed without having to recreate the set each time.

Exporting the Composer Set to PDF

Once the sheets are organized, you can export them to PDF following these steps:

1. Choose File > Export PDF to create a PDF document from your selected sheets.

2. On the Export PDF dialog select All for the Plot Range (Figure 6.21).

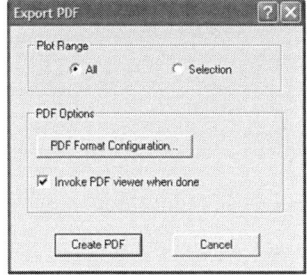

Figure 6.21 MicroStation PDF Composer Export PDF dialog

3. Select the PDF Format Configuration button to display the PDF Format Configuration dialog so you can set the PDF options needed (Figure 6.22).

4. Check Include URLs/Engineering Links, Searchable Text and Levels/Files (Optional Content) on the PDF Format Configuration dialog Format Properties tab.

Note: You can also add your own DWG files or the DWG file used earlier in this chapter. You will need to set the plot area in the Properties dialog - Area Tab.

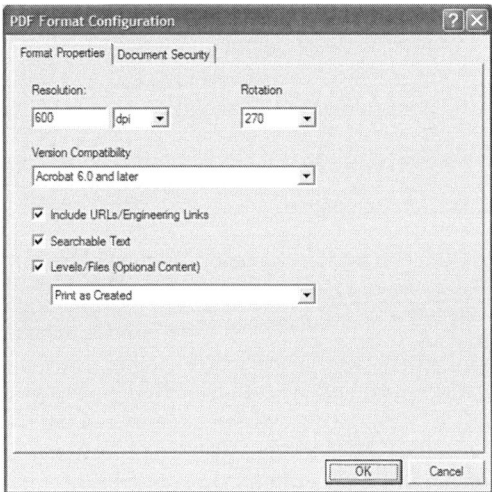

Figure 6.22 PDF Format Configuration Format Properties dialog

Tip: If you always want all the PDF content to be printed even when level/file information is turned off in your PDF viewer, select Print as Created. Print as Displayed will print only the content that is shown on the screen.

Adding Security

Protect the content of the drawing using passwords by following these steps:

1. Select the Document Security tab and check Require a password to open the document and type in a password which will be used to open the PDF document (Figure 6.23).

2. Check Use a password to restrict printing and changes and type in a password control printing and changes.

3. Click OK on the PDF Format Configuration dialog to apply your changes.

4. Check Invoke PDF viewer when done on the Export PDF dialog so that your PDF viewer opens after the PDF is created.

5. Select the Create PDF button and select an output folder and file name for your PDF document.

Select None in the Printing Allowed pull-down to restrict viewers of the PDF from printing. Select Filling in form fields and signing to restrict viewers of the PDF from making any other changes except filling in form fields and electronically signing the document.

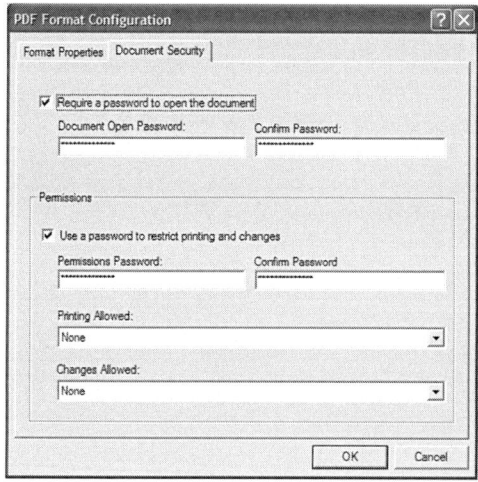

Figure 6.23 PDF Format Configuration Document Security dialog

PDF Navigation

The PDF document includes searchable text and layer information, just as your PDF document did that was created in MicroStation. In your PDF viewer you will see areas in red which are clickable links to follow. Click on Step Details in the upper left of the sheet ar1. The title block area of the sheet ar1 has a external link to a web page, attached document link, and a digital signature field.

Open the bookmarks tab and you will note that there are sub-bookmarks that can be clicked on to navigate to areas in the PDF. Attached documents are denoted by a push pin – these documents are included in the PDF you created.

Note: MicroStation PDF Composer delivers a tutorial that displays the first time you open the product. You should run though this tutorial to learn more about how to create interactive Adobe PDFs.

Universal 3D (U3D) – Jo Terri Wright, Bentley Systems

3D engineering drawings are becoming more important as computers have become able to handle 3D rendering of products. Using 3D drawings will greatly reduce the product development timetable, as well as requirements for field modification of projects due to conflicts from 2D designs.

A consortium of 30 companies, including Intel, Bentley Systems, and Adobe, worked together as the 3D Industry Forum to create a standard for 3D renderings called U3D [1]. ECMA (originally European Computer Manufacturers Association, now ECMA International) has adopted the Universal 3-D First Edition as ECMA-363 [2].

Complex 3D renderings can be shared with anyone using Adobe Acrobat 7 Professional and Adobe Reader 7, as both programs support U3D. The ability to include rich, interactive 3D models in a PDF document allows even novices to navigate a 3D rendering the way the designer intended.

MicroStation 3D to PDF

Bentley Systems offers several ways of publishing U3D content from its design files. From MicroStation V8 2004 Edition, users can choose to export U3D files by choosing File > Export > U3D. Once exported, insert the content into a PDF file using Adobe Acrobat 7 Professional's 3D tool. The Acrobat tool is helpful if you have a presentation document that needs a 3D model included for presentation purposes, for example.

MicroStation's Print interface can publish PDF documents that include the U3D file, eliminating the two-step process. If you are working in a 3D design file and wish to publish a PDF that only contains the 3D model, all you need to do is check off the Plot to 3D option on the print dialog as shown in Figure 6.24.

Of more importance is the ability to publish your 3D models along with your 2D designs by selecting the Plot to 3D option in the Reference Attachment Settings dialog (Figure 6.25).

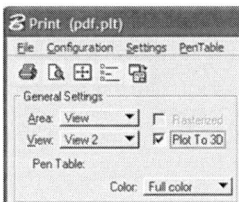

Figure 6.24 MicroStation Print dialog, Plot to 3D

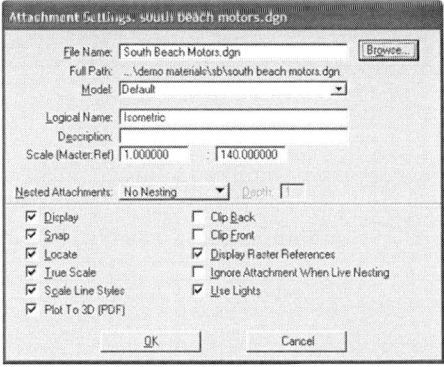

Figure 6.25 MicroStation Reference Attachment Settings dialog, Plot to 3D (PDF) option

The model will appear in the PDF as a U3D on the 2D sheet when published from MicroStation, MicroStation PDF Composer or the ProjectWise publishing products.

Figure 6.26 shows a resulting PDF created with MicroStation PDF Composer when a 3D reference is attached to a sheet. There are controls for interacting with the model, such as Rotate, Walk, Pan, and Zoom.

If animations were associated with the 3D model in MicroStation, those animations will be included in the U3D file and can be played by selecting the Play Animation option. Saved views created in MicroStation are also included in the U3D file and can be chosen by selecting the Manage Views pull-down. Playing the animations and changing views can be controlled by linking to the associated java script with the U3D. These links can be from a bookmark or from the page itself. All that is needed to view and navigate a PDF with a U3D model is the free Adobe Reader 7.

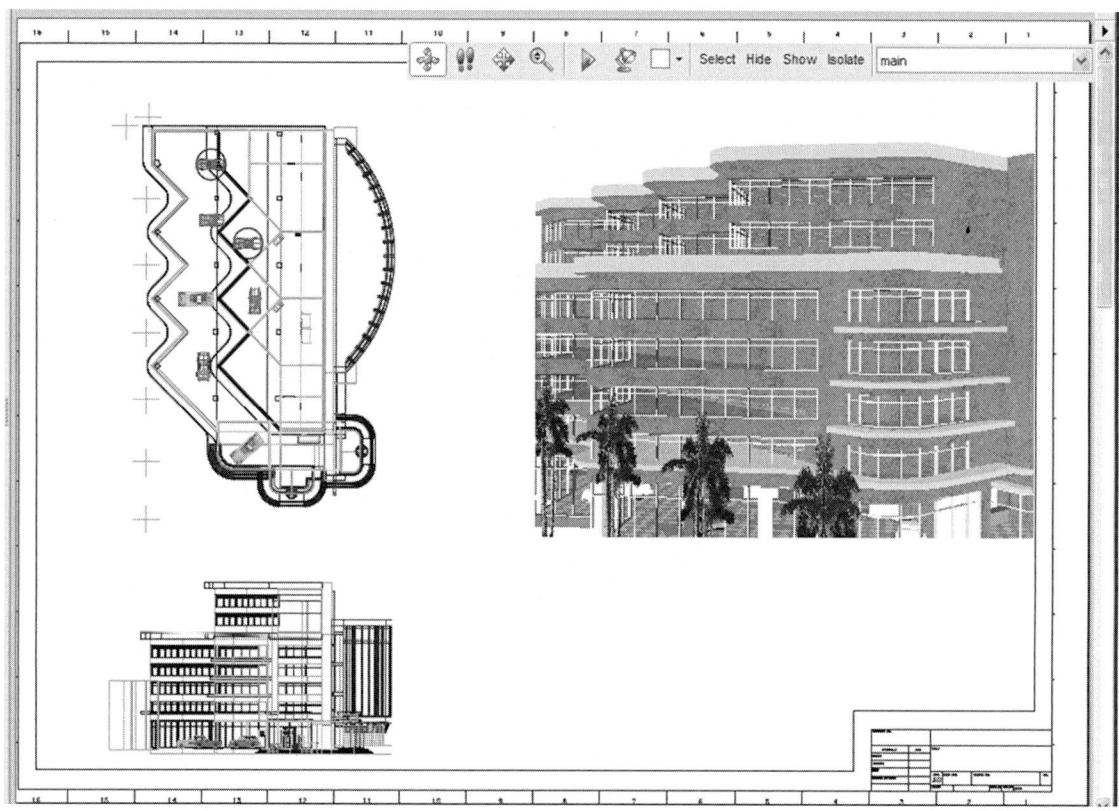

Figure 6.26 PDF document with an attached U3D file created from MicroStation PDF Composer

Currently Bentley Systems' MicroStation and ProjectWise are the only products publishing U3D directly into Acrobat. In a short period there will be numerous programs supporting the effort. It will not be long until your riding lawnmower will come with an Electronic Owner's Manual that shows in 3D how to replace the belts!

Viewing Layered PDF Files in Acrobat

A layered file displays unique features when opened in Acrobat, as you have seen in both this chapter and in Chapter 4. When a layered document is opened in Acrobat, a Document Status dialog box opens and describes the unique features of the file. You will see information about the layers; and if you have included Object Data capabilities, information about the data is also listed (Figure 6.27).

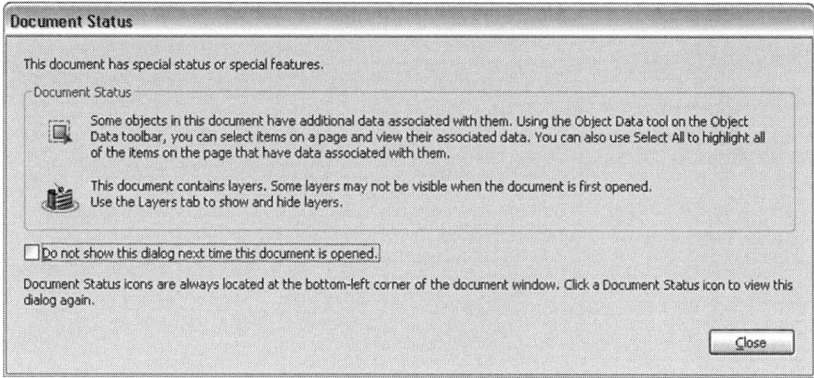

Figure 6.27 Layers and object data are described

After the document is opened, icons display at the left of the Status bar identifying the special features, such as Object Data and Layers icons ![icons]. Layers are listed in the Layers pane, which may or may not be open depending on the features assigned to the document before saving it initially. If you cannot see layers, choose View > Navigation Tabs > Layers to open the pane.

Depending on rights granted by the author of the document, you may be able to modify some of the layer's characteristics. Right-click a layer on the Layers pane and choose Layer Properties from the shortcut menu. The Layer Properties dialog box shows information about the layer, such as its name, visibility, and print and export statuses (Figure 6.28).

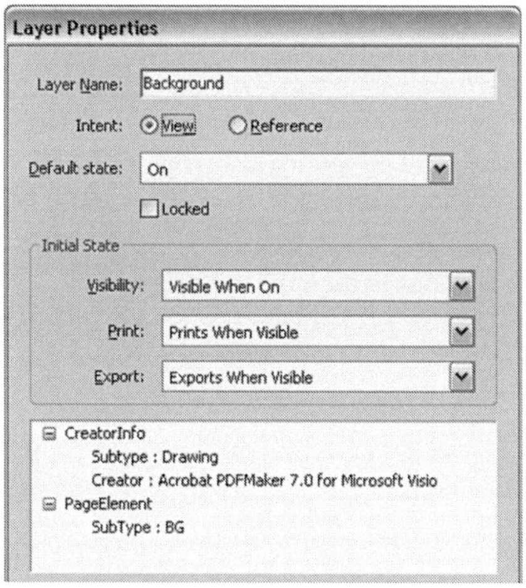

Figure 6.28 Read about a layer's properties

In some circumstances you may have a layered document that contains a static background layer. In Acrobat, any layers imported into the file are called *View* layers, as their purpose, or *intent*, is to display specific content in the drawing. A layer used as a background layer in the source file is also listed as a View layer in Acrobat.

To manage a background in Acrobat, you can click Locked on the Layer Properties dialog box to prevent users from toggling the layer's visibility off.

You can also modify the intent of the layer by following these steps:

1. Right-click the layer on the Layers pane and click Properties to open the Layer Properties dialog box.

2. Click the Reference Intent radio button; the Default State and Initial State settings are disabled.

3. Click OK to close the dialog box. In the Layers pane, the modified layer is permanently visible, and no longer has a user-controlled intent. Moving the mouse over the listing displays a tooltip stating the layer is a reference layer (Figure 6.29).

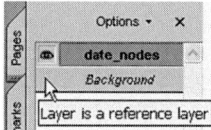

Figure 6.29 Use a reference layer

Summary

In this chapter we looked at converting AEC file formats to Acrobat PDF files, and saw ways to establish the layer structure of the files. AutoCAD contains a PDFMaker which can be used in a similar way to Microsoft Office program's PDFMakers for converting files. Layers can be managed and defined as layer sets for future conversions. A converted drawing is used as a source document for conducting a review, and the results of the review can be integrated back into the source AutoCAD drawing.

Bentley Systems' MicroStation and ProjectWise product lines can publish PDF documents from DGN and DWG files, including vector, raster, and 3D model data. The programs include processes for printing and batch printing that can be configured and reused. Converted drawings can retain links and bookmarks, searchable text, 3D models, bookmarks, and password security.

Acrobat offers special features for managing and viewing layers in Acrobat using the Layers pane and layer properties.

Exercises

1. Using the methods outlined in the chapter, convert sample AutoCAD and MicroStation documents (depending on your available programs.)

2. Experiment with exporting layer sets and subsets of drawings.

3. Save a PDF settings file from the PDFMaker and then reuse the file in a subsequent conversion process.

4. Examine a layered document in Acrobat, either one you have converted or one of the sample files.

Copyright Notice: Many elements of this project have been modified due to copyright and liability issues, and should not be used in other projects. They are only being used to teach the processes in the context of this book.

Project

The goal in this project is to create two PDF files from AutoCAD drawings that are added to the project in the following chapter – use the source material in the **ch06_project** folder.

Task 1: Converting an AutoCAD drawing to PDF using the PDFMaker

1. Open the AutoCAD file **022709_C05.dwg** in AutoCAD.
2. Choose Adobe PDF > Convert to PDF.
3. Choose whether to flatten or maintain layers (the project is not dependent on either option.)
4. Work with AutoCAD to print the file to D size paper.
5. Save the file as **022709_C05.PDF**.

Task 2: Printing an AutoCAD drawing to PDF

1. Open **022709_C07.dwg** in AutoCAD.
2. Choose File > Print and select the Adobe PDF printer driver. Make sure the drawing is printed to the PDF format using the D-Size drawing size.
3. Click Save to print the file as **022709_C07.PDF**.

References

[1] The first edition of the specification is now available from 3DIF, *3DIF, The 3D Industry Forum,* n.d. http://www.3DIF.org.
[2] The first edition of the specification is available from ECMA-International, *TC43 - Universal 3D* (U3D) n.d. http://www.ecma-international.org/memento/TC43.htm.

7

Assembling the Package

Content from a number of documents can be combined in Acrobat to make one document using either the binder process, described in Chapter 5, or by manipulating document pages. A common workflow is to create a collection of PDF files that can then be combined and recombined in Acrobat to make a cohesive, finished file.

Acrobat also contains features that further enhance the unity of a document, including a range of elements that can be added to a PDF file. Keeping track of large numbers of PDF files can be a very difficult task; this task is made much simpler with the introduction of the Organizer window in Acrobat 7.

In this Chapter

In this chapter you learn ways to manipulate the page content in a file and how to manage and organize PDF files and attachments.

You will learn about:

- Working with different document commands to add, remove, replace, and otherwise manipulate page content in a file
- Cropping and resizing pages in a file
- Using the Pages pane to navigate content in a file
- Adding page numbers to a document
- Applying header/footer elements to a document
- Using backgrounds and watermarks in a PDF document
- Organizing, sorting, and accessing files using Acrobat's Organizer window

Manipulating Pages

Rather than working through the Create PDF task button's command to create a document from multiple files, you can manipulate pages manually using the Document menu's commands and the Pages pane, one of the default Navigation tabs in Acrobat. If the pane is not displayed, choose View > Navigation Tabs > Pages and dock the tab with the other navigation tabs at the left of the program window.

There are a number of locations in which to select the same commands in Acrobat. Commands may be chosen from the Document menu, the Options menu on the Pages pane, or by right-clicking a page thumbnail to display the shortcut menu in the Pages pane. In the procedural descriptions that follow, the command is listed from the Document menu.

In all descriptions, the pages being manipulated are selected first – using this process saves time, as the dialog box corresponding to the selected command lists the selected page or pages automatically. In each case, you can also replace the designated pages by typing different values in the fields. Acrobat includes pairs of parallel commands for deleting/inserting pages, as well as extracting/replacing pages. In the final document, the number of pages may be the same, but there are different reasons for using one command versus the other, as described in this section.

Deleting Pages

To delete one or more pages from a file, open the document and click the Pages tab to display the Pages pane. Follow these steps:

1. Click the thumbnail for the page you want to remove in the Pages pane. The selected page is displayed in the Document pane, and the thumbnail is highlighted in the Pages pane.

2. Choose Document > Delete Pages.

3. The Delete Pages dialog box opens and the selected page is listed automatically.

4. Click OK to close the dialog box; click OK again to close the confirmation dialog box. The selected page or pages are deleted.

To bypass the dialog boxes for faster deletions, select the page or pages in the Pages pane and press Delete; click OK in the confirmation dialog box.

Inserting Pages

Inserting pages is a similar process to deleting. You can immediately insert the content from a PDF document into an open PDF file; if the content to be inserted is in another format, Acrobat converts it to PDF first, and then inserts it into the subject file.

Open the file into which you want to add pages, and follow these steps:

1. Click the Pages tab to display the Pages pane, and then click the thumbnail for the page prior to the location where you want to insert additional pages.

2. Choose Document > Insert Pages to display the Select File To Insert dialog box. Locate and select the file for insertion and click Select. The dialog box closes, and the Insert Pages dialog box opens.

3. Specify the insertion location in the document; a selected thumbnail's page number is shown automatically in the dialog box.

4. Click OK to close the Insert Pages dialog. Acrobat adds the page or pages to your document.

Inserting Pages Visually

When you are working with highly distinctive pages, such as slides or large images, you can also manipulate the content visually in the Pages pane (Figure 7.1).

Open the subject documents and choose Window > Tile > Horizontally (or Vertically) to tile the windows. Open the Pages panes on each document to display the thumbnails. Select thumbnails from one document and drag them to the other document. You see a vertical bar displayed on the Pages pane showing the insertion location.

Extracting Content

To quickly create several separate documents from a single document, you can use Acrobat's Page Extraction feature. After selecting the page or pages to extract in the Pages pane, choose Document > Extract Pages to open the Extract Pages dialog box, listing the selected page or pages. Select an extraction option and click OK to close the dialog box and extract the content (Figure 7.2).

There are three ways in which to perform an extraction. The default method is to create a separate document from the selected pages, without affecting the original document's contents. You can also choose Delete Pages After Extraction, which removes the selected content from the original document permanently, and places it in a unique document. For either of these methods, the extracted content is named with "Pages from" prepended to the original file's name.

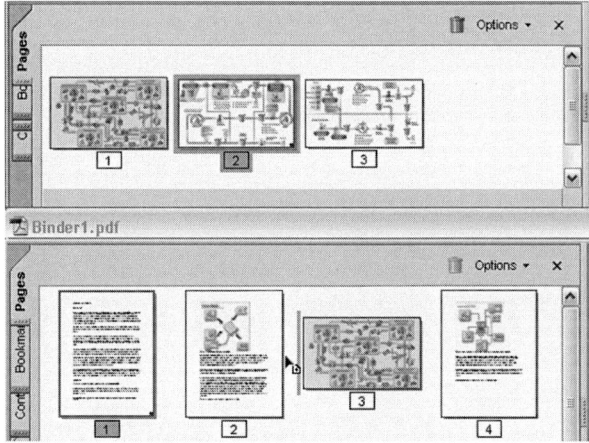

Figure 7.1 Visually insert pages from one document into another

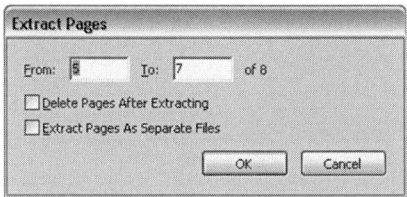

Figure 7.2 Choose a method of extraction in the dialog box

Alternatively, you can Extract Pages As Separate Files to create a PDF file for each page specified in the dialog box. When you choose this method, first select a storage location in the Browse for Folder dialog box. After extraction, each extracted page is named using the original file's name with the page number appended to it and saved in the folder specified.

If you want to extract noncontiguous pages, first reorder the pages so the extracted content is listed consecutively, and then proceed with the extraction.

Replacing Pages

Replacing pages may seem, on the surface, the same as inserting pages. The final outcome is the same; that is, in either case you have substituted content in the file. However, the two actions have quite different effects on the receiving document.

If you have added content to a page, such as links or comments, deleting and replacing the page also removes the additional content. Instead, if you use the Replace Pages command, only the underlying page is replaced, leaving the additional content intact.

Follow these steps to replace pages in a file:

1. Select the thumbnails for the page or pages in the Pages pane, and choose Document > Replace Pages.

2. In the resulting dialog box, locate the document containing the pages you want to use for the replacement and click Select. The dialog box closes, and the Replace Pages dialog box opens listing the name of the selected file.

3. Specify page numbers in both the Original and Replacement sections of the dialog; selected pages in the original document are identified in the Original pages fields.

4. Click OK, and then click OK again to confirm the replacement to close the dialog box and make the modification to the file.

Configuring the Pages Pane and Thumbnails

Increase or decrease the size of thumbnails in the Pages pane by choosing Enlarge (or Reduce) Page Thumbnails from the Page pane's Options menu. Decrease the size of the thumbnails for an overview of a document's content; increase the size of the thumbnails when using large images or slides to see the content more easily. Resize the Pages pane to display the thumbnails in multiple panels if necessary by dragging the divider bar at the right edge of the Pages pane right to increase the width of the pane.

Thumbnails can be embedded or unembedded. Embedded thumbnails make viewing a large document faster, as the thumbnails do not have to be redrawn. However, embedding adds to file size. When changes are made to the document the thumbnails are not automatically updated, but must be un-embedded and re-embedded using the appropriate commands on the Options menu of the Pages pane.

Cropping Pages

Click the Crop tool [🔲] on the Advanced Editing toolbar, and drag a marquee roughly the size of the area you want to crop on the page on the Document pane. Double-click within the marquee to open the Crop Pages dialog box and show the marquee in the dialog box's preview area.

Cropping can be applied to an entire document or to selected pages by following these steps:

1. Choose Document > Pages > Crop to open the Crop Pages dialog box; the Crop radio button is selected by default (Figure 7.3).

2. Click the Units drop-down arrow to select an alternative unit of measure for the cropping rather than the default inches if necessary.

3. Click Constrain Proportions to crop the page equally on all four sides, or adjust the individual margins using the Margin Controls by typing new values or clicking the arrows for each margin to reset its value. As you make adjustments, you can see the cropping outline in the preview area of the dialog and also on the Document pane adjust. Click Set to Zero to restore the original margins of the page.

4. Instead of manually adjusting margins, choose a page size from the Page size drop-down list, or type height and width values for Custom page size. Choosing a page size disables the margin settings (Figure 7.4).

5. Specify the cropping page range. Pages preselected in the Pages pane are automatically shown in the fields. Cropping can apply to all pages, or only even or odd pages.

6. Click OK to close the dialog box and apply the cropping. Save the document.

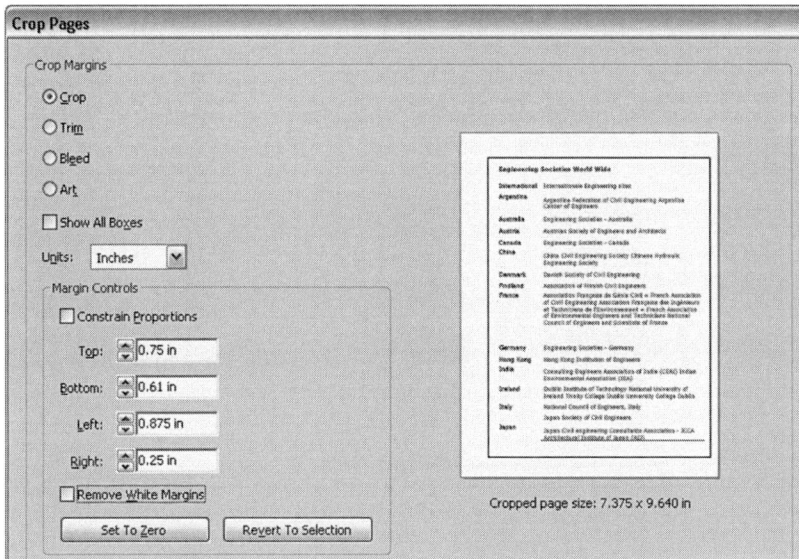

Figure 7.3 Adjust margins and apply cropping

You cannot undo cropping by choosing Edit > Undo. If necessary, choose File > Revert to return to the uncropped version.

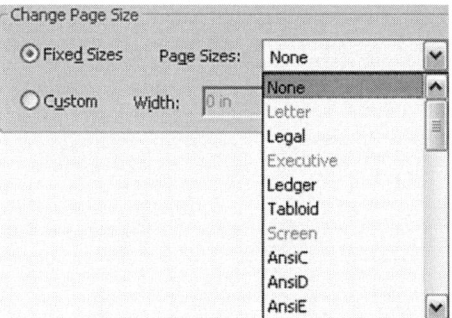

Figure 7.4 Choose a preconfigured or custom page size

Note: In addition to other page manipulations, you can also rotate pages. Choose Document > Pages > Rotate to open the dialog box; specify the page or pages, the angle of rotation, and the orientation.

Numbering Pages

All PDF documents show page numbers on the Status bar below the Document pane. Pages are numbered in logical order, but you can modify the pagination system using number prefixes, sections, and other customizations. If you anticipate using documents within a collection or combined with other content and plan to use page numbering in Acrobat, remove the page numbers (and other footer content) before creating the PDF files for convenience.

Follow these steps to apply custom section numbering to a document:

1. Select the thumbnails for the pages in the first section in the Pages pane, and choose Number Pages from the Pages pane's Options menu to open the Page Numbering dialog box. The Selected option is active if pages are selected before opening the dialog box and pages are named in the From fields (Figure 7.5).

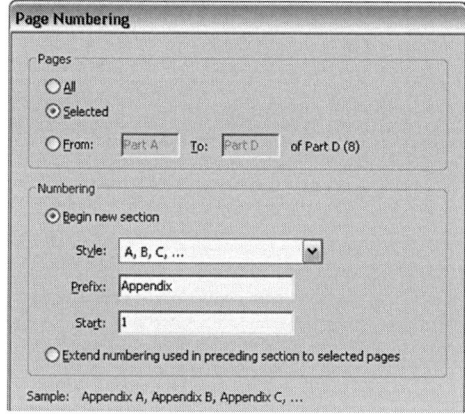

Figure 7.5 Specify page numbering, including custom sections

2. Click the Style drop-down list to choose a page format in the Begin new section area.

3. Type a value and punctuation in the Prefix field. The page numbering starts at "1" by default, as shown in the Start field, shown in Figure 7.5.

4. Click OK to close the dialog box and modify the document's page numbers.

Figure 7.6 shows the document's renumbering using the settings shown in the previous figure. The selected pages are now numbered as Appendix A through Appendix D.

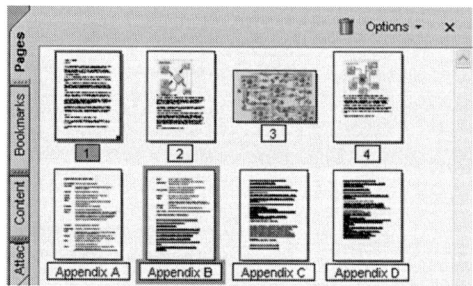

Figure 7.6 Pages are repaginated using the custom page numbering

Both the custom page numbering and the total page count are displayed in the status bar below the Document pane (Figure 7.7). To modify existing numbering, select the pages and then choose Option > Page Numbering to reopen the Page Numbering dialog box. Click the Extend numbering used in preceding section to selected pages radio button, and click OK to close the dialog box. The additional pages are numbered using the pre-existing numbering format. If you want to modify labels, such as chapter numbers, select the pages and reopen the Page Numbering dialog box to make changes.

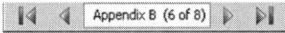

Figure 7.7 The Status bar shows both custom and default pagination

Applying Headers and Footers

Adding common headers and footers is a good method of creating a visually unified document, regardless of where the components originated. If you are planning to add headers and footers in Acrobat, remove any pre-existing header or footer content in the source files before converting to PDF to save time.

You apply headers or footers using the same process using the associated tabs in the Headers & Footers dialog box. Follow these steps to add a footer to a document:

1. Choose Document > Add Headers & Footers to open the Add Headers & Footers dialog box.

2. Click the Footer tab; you see three text boxes at the top of the dialog box used to display left-justified, centered, or right-justified content (Figure 7.8).

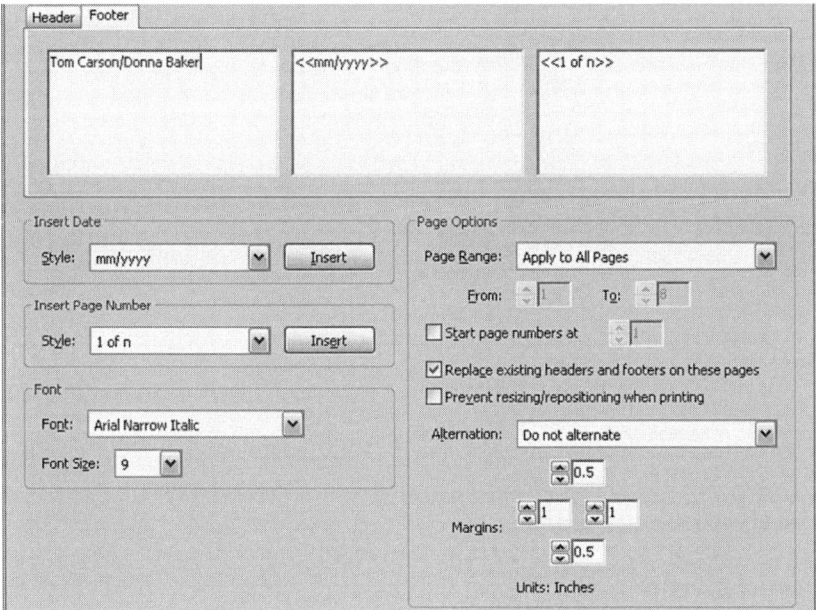

Figure 7.8 Configure the content and layout of headers and/or footers in the dialog box.

3. Choose the font and font size from the drop-down lists below the header/footer content boxes, which are only shown on the preview and Document pane.

4. Choose other content to insert such as dates and page numbers, choose styles, and enter text.

5. Click the text box in which you want to insert the page numbers or date, and click Insert. You can add custom text to a text box by clicking to activate a text cursor and then typing.

6. In the Page Options section, choose page ranges, numbering, margins, and whether to use even or odd pages only. Adjust margins using the arrows, which increment or decrement in one-half inch amounts, or type a value in the fields.

7. Click Preview to see the layout of the footer elements in a pop-up Preview window; you won't see a custom font unless the font size is large enough. Click OK to close the Preview window.

8. Click OK to close the dialog box and apply the footer. The content and numbering appear on the page in the specified areas (Figure 7.9).

If you need to make any changes, reopen the dialog box, make the modifications, and click Replace existing headers and footers on these pages. Click OK to close the dialog box and make the changes.

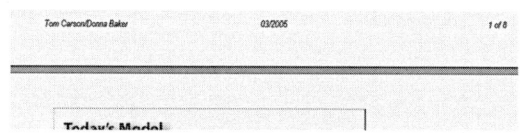

Figure 7.9 Use headers or footers to unify a document

Adding Watermarks and Backgrounds

You can add backgrounds behind the content on a document's pages or overlay the content using a watermark. In Acrobat, both watermarks and backgrounds may be either text or graphics. You can configure text directly in the dialog box, or import an image. Open the document, and follow these steps to apply a text watermark:

1. Choose Document > Add Watermark & Background to open the dialog box (Figure 7.10).

2. Select a watermark or background option and specify where you want the item to display. Watermarks and backgrounds are shown both on screen and in print by default.

3. Click the From Text radio button to activate the text options. Type the text in the text field, and choose font characteristics. If you want to use a font size that fits the page, it is simpler to leave the default font size and click Fit to page in the Scale options on the dialog box.

4. Define position and appearance characteristics; specify rotation and opacity if desired.

5. Select the Page Range options to specify where the background or watermark is applied.

6. When the watermark is configured correctly, as shown in the preview area, click OK to close the dialog box and apply the watermark.

Using a Graphic Image

A graphic image in PDF, BMP, or JPEG format is imported into the subject document. To use a graphic image as a background/watermark, follow these steps:

1. Open the document and then open the Add Watermark & Background dialog box.

2. Click From file in the Source section of the Add Watermark & Background dialog box, and click Browse to locate and select a source file for the background/watermark. Select the page if the source document contains more than one page.

3. The selected image displays in the preview area (Figure 7.11). Configure other settings as required and click OK to close the dialog box and apply the background/watermark.

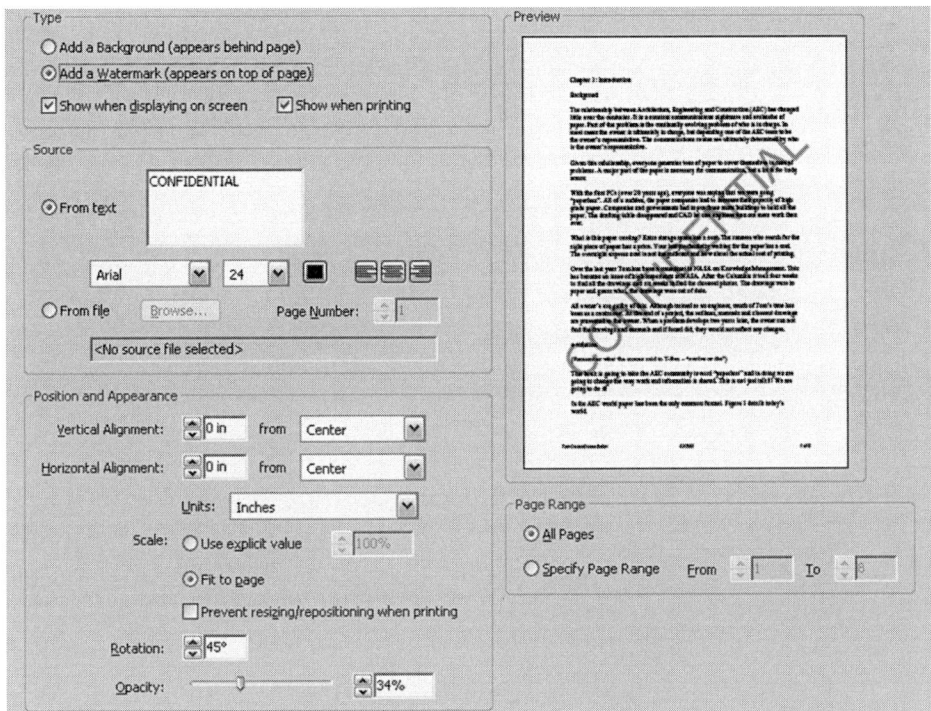

Figuro 7.10 Dcfinc content and characleristics for a watermark or background

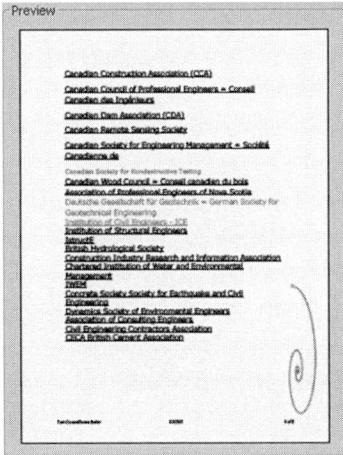

Figure 7.11 Apply a graphic image as a background or watermark.

Organizing PDF Files

New in Acrobat 7 is an Organizer, a separate window used to manage and view PDF files on your computer. To access the Organizer, click the Organizer button 📁 on the File toolbar or choose File > Organizer > Open Organizer. The Organizer contains three frames (Figure 7.12). Resize the frames in the window by dragging the splitter bars.

The Organizer also includes several commands you can use in addition to organizing PDF files, such as opening, printing, or emailing a file. You can also create a binder using the command from the toolbar, or start a review cycle.

Click a file or folder in the Categories pane at the left of the window to display its contents and thumbnails in a list in the central Files pane; click a file in the Files pane to display its contents in the right Pages pane.

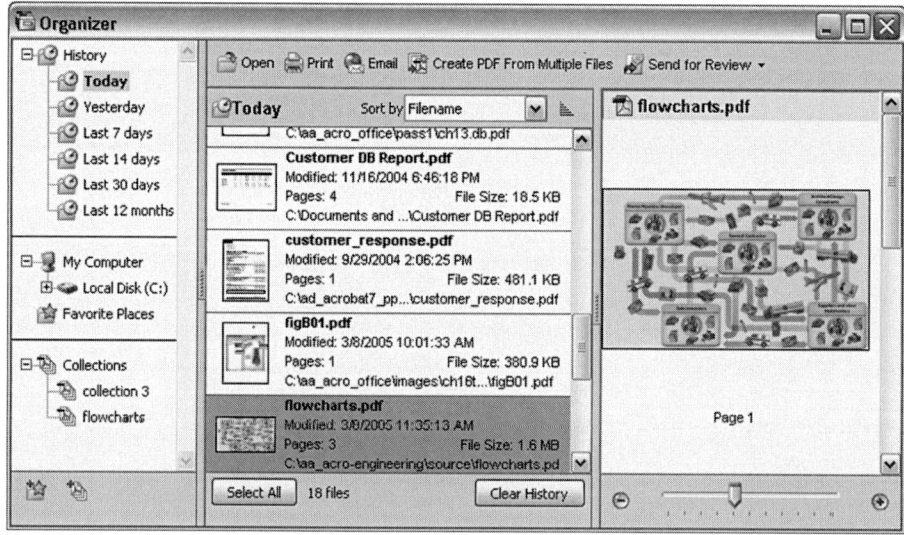

Figure 7.12 Manage and control PDF files in the Organizer window

Using Backgrounds and Watermarks

Here are some tips for working with backgrounds or watermarks:

- The Add Watermarks & Backgrounds dialog box uses the first page of the document as a sample. To show another page instead, move that page to the start of the document before opening the dialog box.
- When a background or watermark is applied and saved, it cannot be removed. This is to preserve the document's integrity. If you aren't sure about using the feature, save a copy of the file without the added content.
- If you do not have a copy of a file and decide you do not want to use the watermark or background, open a blank document in a source program, such as Word, and save it as a PDF. Then apply it to the document as an image in the Add Watermarks & Backgrounds dialog box.
- Instead of processing a large number of images in an image editing program to apply a watermark, combine them into one PDF document and then add a text or image watermark in Acrobat.

Categorizing Files

The Categories pane uses a hierarchy of folders. The categories inclue History, folders on your computer, and Collections, all are used to organize and manage the PDF files on your computer.

History

The Organizer's History works the same way as a Web browser's history. There are several time frames to choose; click a time frame in the History pane to show the list of PDF files opened within that time frame in the Files pane.

To delete a listing, select a time frame and click Clear History at the bottom of the Files pane. Be careful when deleting listings, as clearing a history also clears all histories of shorter duration. For example, clearing the Last 14 days history also deletes the Last 7 days, Yesterday, and Today listings.

Favorite Places

You can include folders or files in the Favorite Places heading in the My Computer listings. Click Favorite Place 🏛 at the bottom of the Categories pane to open a Browse dialog; locate and select the folder or file and click OK. The selected folder is added to the Favorite Places category.

Collections

A Collection functions as a virtual folder. It is not a folder constructed within your operating system, but is used only within Acrobat for organizing content. Manipulating content in a Collection has no impact on the files or folders on your computer. A common use of the Collection method of organizing files is to streamline project management.

Acrobat's default installation includes three blank collections, named *collection 1* through *collection 3*. To use an existing collection, right-click its name in the Categories pane and select an option from the shortcut menu (Figure 7.13). You can rename, delete, or add files to the selected collection. To list files, choose Add Files from the shortcut menu, and locate and select files through the resulting dialog box. Click Add to close the dialog box and add files to the collection.

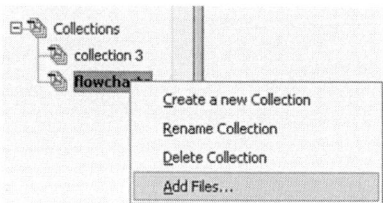

Figure 7.13 Organize files in Collections for easy access

To add another collection, right-click the Collections heading in the Categories pane or an existing collection's label and choose Create a new Collection from the shortcut menu, or click the Collections icon 🏛 at the lower left of the Organizer window.

Build and delete collection folders when working on a project for easy access to working files. To organize very complex projects, you can nest collections within collections. Select an existing collection and choose Add Collection from the shortcut menu to nest a sublevel folder.

Listing Files

Any selection you make in the Category pane is listed in the central Files pane of the Organizer. Unless a document is protected by a security policy, you will see both a thumbnail and information; a secured document shows only a PDF document icon (Figure 7.14).

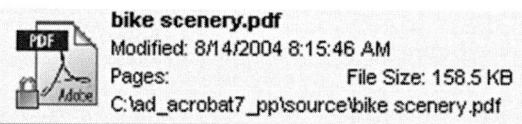

Figure 7.14 A secured document displays only an identifying icon

The default display lists documents in alphabetical order; the selected category is shown at the top of the pane. Click the Sort by drop-down arrow and choose an alternate sort method from the list (Figure 7.15).

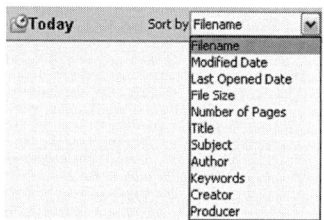

Figure 7.15 Sort the contents of a selected category

Viewing Page Previews

A file selected in the Files pane displays its content, provided it is not a protected file, in the Pages pane. Use the scrollbars to view a page or view different pages in a multi-page file; resize the display using the zoom slider or buttons below the preview (Figure 7.16). You can resize the display of the preview using the zoom slider or buttons below the preview area, and use the vertical scrollbars for viewing the page. Double-click the page preview to open the file in Acrobat displaying the page shown in the preview.

Using the Organizer's Features

You can access the History and Collections from Acrobat instead of opening the Organizer window:
- Choose File > History and a date option to view the same listing as that shown in the Organizer, or choose File > Organizer > Collections to view collection contents.
- Click the Organizer icon's drop-down arrow on the Basic toolbar to display its menu and choose a file from the history listings.
- Click the Organizer icon's drop-down arrow to display its menu and choose your collection.

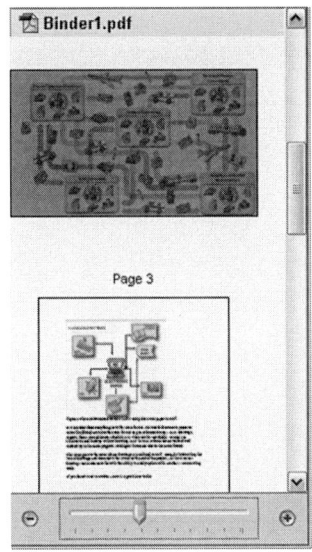

Figure 7.16 Preview file content before opening a document

Summary

In this chapter we looked at different ways to manipulate document contents. You saw a number of ways to change the characteristics of the pages in a PDF file by adding and removing content, or substituting existing content. You learned how to add unifying elements to a document such as page numbers, as well as headers/footers and watermarks/backgrounds. Finally, you were introduced to Acrobat's new Organizer window, used for collecting and managing PDF files.

Exercises

1. Using the methods outlined in the chapter, manipulate the content of sample documents by adding and deleting pages. Extract content from an existing document using the different options listed in the chapter; replace content in a document using the method described in the chapter.
2. Add and configure page numbers in a sample document.
3. Add a header and a footer to a sample document.
4. Add and configure a text watermark/bookmark for a sample document.
5. Add graphic bookmark/watermark to a sample document.
6. Create and manage a set of collections in the Organizer.

Copyright Notice: Many elements of this project have been modified due to copyright and liability issues, and should not be used in other projects. They are only being used to teach the processes in the context of this book.

Project

The goal in this project is to manipulate the file content for the project and complete the assembly of the project pages. The tasks include organizing, inserting, rotating, and cropping the pages. Since the drawings are not for construction as they are not printed to scale, they need to be labeled as such using a watermark. You will add a header, and create a collection in the Organizer to store the project files. Use the PDF project file from Chapter 5's project, as well as the two PDF files created in Chapter 6's project. If you do not have access to AutoCAD or MicroStation, PDF versions of the two CAD drawings required in this project are available in the **ch07_project** folder.

Task 1: Organizing project pages

1. Open the **DuPont Soccer Park Civil.pdf** file in the ch07_project folder.
2. Click the Pages tab to display the Pages pane.
3. Drag the thumbnails into the proper order to rearrange the document's pages. Look for the page numbers at the lower right of each drawing – **Drawing 02279_C01-Model.pdf** (shown as 2 in the Pages pane) should be the first page.
4. Drawing **02279_C04-Model.pdf** is inverted. To correct its orientation, right-click the thumbnail in the Pages pane (shown as 4) and choose Rotate Pages > 180 degrees.

Task 2: Inserting project pages

1. Insert another page. Click **Drawing 02279_C01-Model.pdf** and choose Document > Insert Pages. Follow the dialogs and insert **Drawing 02279_C02-Model.pdf** to follow page 1.
2. Using your desired insertion method, insert your converted PDF files for the two additional drawing pages to follow page 4. The files were created in Chapter 6's project. If you do not have access to AutoCAD or Microstation, you can insert the files supplied in the ch07_project folder, named **02279_C05-Model If you do not HAVE CAD.pdf** and **02279_C07-Model If you do not Have CAD.pdf**.

Task 3: Sizing project pages

1. Select the Crop tool from the Advanced Editing toolbar. Drag a marquee around the border of page 1 and release the mouse.
2. Double-click the box with the Crop tool to open the Crop Pages dialog box.
3. Adjust the margins numerically on the dialog box if required.
4. Click the All button in the Page Ranges area on the dialog box to apply the cropping to all drawings.
5. Click OK to close the dialog box and initiate the page crops.
6. Check the pages individually in the project to evaluate the cropping. To make adjustments to any margins, double-click the drawing with the Crop tool to reopen the Crop Pages dialog box and adjust the margin settings. Click OK to close the dialog box and adjust the margins on the page.

Task 4: Applying a watermark

1. Choose Document > Add Watermark & Background to open the dialog box.
2. Click the Add a Watermark radio button at the top of the dialog box, and leave the Show when displaying on screen and Show when printing check marks selected.

3. Click the From text radio button in the Source area and type "Not Released for Construction" in the From text field.
4. In the Position and Appearance section, click Fit to Page to size the text automatically.
5. Click the Rotation field and type 45 to rotate the text.
6. Drag the opacity slider to 20%.
7. In the Page Range area, make sure the All pages radio button is selected.
8. Click OK to close the dialog box and apply the watermark. Reopen the dialog to make any adjustments necessary; do not save the file unless your watermark displays as desired.

Task 5: Applying a header

1. Choose Document > Add Headers & Footers to open the dialog box, and click the Footers tab.
2. Choose a date and enter text for a header as desired.
3. In the Page Range area, select Apply to All Pages.
4. Click OK to close the dialog box and add the header.
5. Save the file.

Task 6: Building a collection

1. Click the Organizer button on the File toolbar to open the Organizer window.
2. If you are working on these tasks in one session, click Today in the History listings to display the files used in the project to date.
3. Select the files in the Files pane of the Organizer window; you can click Select All to automatically select the set of files.
4. Drag the files to the collection 1 listing (or an empty listing) in the Categories pane to add the files.

8

Modifying Content in the Package

Acrobat is not a document editing program, and although it can use content from a wide range of programs, it is not designed fundamentally as an editor. Fortunately, Acrobat does include a number of tools and features that are used to manipulate aspects of a document's contents, such as text, images, objects, and reading order.

It is common to work with a number of other file formats in addition to PDF files in the same project. Acrobat lets you work with other file formats throughout your workflow, including the ability to export a PDF file from Acrobat in a range of file formats, and attaching different types of file to an existing PDF document.

In this Chapter

In this chapter you see how to modify the content both within and without a PDF file as you learn about:

- Selecting and reusing text in a PDF document
- Extracting tables from a PDF file
- Reusing images in a PDF file
- Touching up objects, text, and reading order
- Saving a PDF document in a range of alternate file format
- Adding attachments to a PDF file as attachments to either the entire file or a selected location
- Managing file attachments
- Using attachments in earlier versions of Acrobat

Selecting Text in a PDF File

As indicated above, Acrobat is not a document editing program, but it does include some tools used for manipulating and editing content in a document. The Select tool, located on the Basic toolbar, is a multi-purpose tool that behaves differently depending on what is being selected on the document. The Select tool on the Basic toolbar is a different tool than the Select Object tool on the Advanced Editing toolbar, which is used for selecting items such as fields or links.

The Select tool can select text, images, and tables. As well as changing in response to the object it is selecting, the tool also provides a menu of options that varies according to the type of object being selected. You may find variations in the allowable actions for selected text. This is not a program error; depending how the document is tagged, and whether or not the content is formatted, you may or may not have access to commands such as Copy with Formatting.

Using the Select tool is an efficient way to select content used for links and bookmarks, discussed in Chapter 9. To select text in a document, follow these steps:

1. Click the Select tool 🔘 on the Basic toolbar and then drag over some of the text you want to select. You see the text highlighted in gray, displaying small arrows at the upper left and lower right of the selection (Figure 8.1).

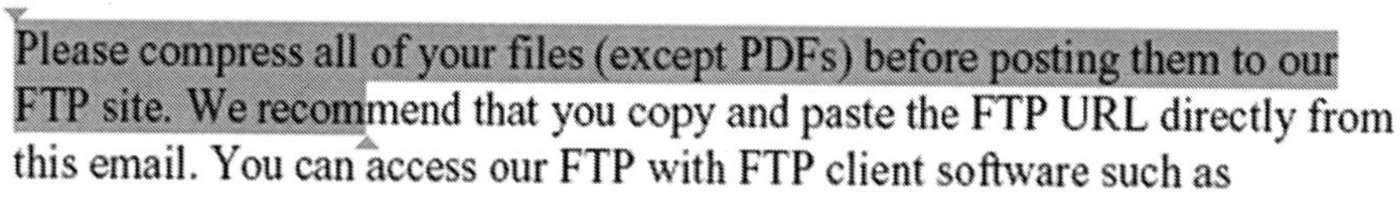

> Please compress all of your files (except PDFs) before posting them to our FTP site. We recommend that you copy and paste the FTP URL directly from this email. You can access our FTP with FTP client software such as

Figure 8.1 Selected text shows a highlight

2. Drag either arrow to add text to the selection.

3. Hold the cursor over the selected text to display the Select Text icon 🔲 .

4. Move the cursor over the Select Text icon to open the associated menu applicable to the selected content and choose a command (Figure 8.2).

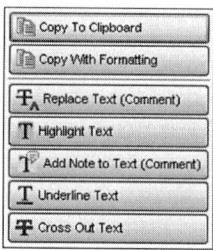

Figure 8.2 Choose commands from the pop-up menu

Selection Options

There are many ways to enhance the efficiency of your text selection methods. Here are some tips:

- Double-click a word to select it; triple-click to select a line of text; quadruple-click to select all the text on a page.
- If you want to select all the text in a document, select some text with the Select tool, then right-click and choose Select All Text from the shortcut menu.
- Add text letter by letter by pressing the Shift and an arrow key in the direction you want to add the text; press Shift+Ctrl and the arrow key to add one word at a time to the selection.
- The Copy With Formatting option is active in a tagged document, and useful for complex documents, such as those containing columns and pull-out quotes.
- In a secured document, you won't be able to copy text if the author has denied copying permission.

- Scanned text that has not been captured is an image only, and cannot be selected (refer to OCR and capture in Chapter 4.)

Selecting and Reusing Images

You can reuse individual images or portions of images from a document, again selecting the content with the Select tool.

Follow these steps to select and repurpose an image in an existing document:

1. Click the Select tool on the Basic toolbar and move the cursor over the subject image on the document. The Select tool changes to the pointer appearance.

2. Click to select the image under your cursor; the image is highlighted. To select part of an image, drag a marquee with the Select tool instead of clicking the entire image to select it.

3. Release the mouse and the Select Image icon displays on top of the selected image (Figure 8.3).

4. Move the cursor over the icon to open the menu, which only allows copying to the Clipboard.

5. Click off the image to deselect it. Once the image or segment is copied to the Clipboard, you can reuse it at will.

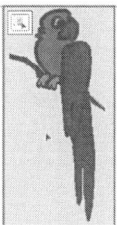

Figure 8.3 The Select tool changes to the Image Select option automatically

An image in a document can be saved as a separate file. Select the image (or drag a marquee to select part of the image). Right-click to open the shortcut menu and choose Save Image As. In the resulting dialog box, name the image file, choose an image format, and specify the save location.

Enabling Selection Preferences

You can automatically enable text selection for the Hand tool by modifying a program preference. Choose Edit > Preferences > General and click Enable text selection for the Hand tool. When you hold the cursor over text in a document, as described in the steps, it automatically converts to the Select tool.

You may have to modify a program preference to select an image with overlaying text. By default, the General Preferences include a setting to select text before images. Choose Edit > Preferences > General and check Select Images Before Text.

Reusing Table Information

It was difficult to deal with tables in PDF files prior to Acrobat 5, which includes a Select Table tool. In Acrobat 6, to reuse a table you export the content as a rich text format (RTF) file, and then reassemble and restructure the table in Word or Excel or another spreadsheet-capable program. In Acrobat 7, selecting table information depends on the tagging status of the document.

In an untagged document, choose the Select tool from the Basic toolbar. Use any of the techniques described earlier in the chapter to select text, and select all or some of the table's contents. A Select Table icon ⁃⃞ displaying when you move the Select tool over the table on the document page indicates the document has a tagged structure (Figure 8.4). The tags include <table> tags, which contain an assortment of table structures (such as rows and cells) as well as content including text objects. Click once to select the entire table.

Hold the cursor over the selected text to display the Select text icon and move the cursor over the icon to show the menu (Figure 8.5). You can also right-click the selected text to open the shortcut menu containing the same commands. In addition to the options available for a table selected in an untagged document, you can also select a Copy with Formatting command if the document is tagged.

Acrobat recognizes the text as belonging to a table format, which gives you three table-specific commands, including saving the selected table as a separate file, opening the table in a spreadsheet, or copying the table to the clipboard for pasting into another file. In both Word and Excel, the tables taken from the PDF document are editable and ready to use, as Acrobat exports content using a comma-separated value (CSV) structure.

Travel	Lodging	Meals
35.00	669.46	225.00
300.00	250.00	270.00
60.00		100.00

Figure 8.4 Select some or all of the text in a table

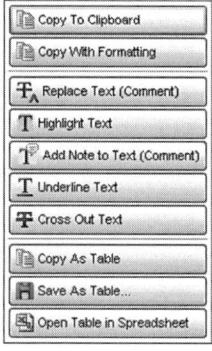

- Copy To Clipboard
- Copy With Formatting
- Replace Text (Comment)
- Highlight Text
- Add Note to Text (Comment)
- Underline Text
- Cross Out Text
- Copy As Table
- Save As Table...
- Open Table in Spreadsheet

Figure 8.5 Choose from a range of commands to apply to the selected table content

Note: If you want to copy and paste a table from a tagged file, select the content through the Tags pane for easy selection. Open the Tags pane and click the table's <table> tag. Then choose Options > Copy Contents to Clipboard from the Tags pane's menu to copy the table's content and its formatting.

Editing Text in a PDF

Sometimes you find the odd typo or two in a PDF file in spite of your best intentions, or decide that you would like to rephrase some text. Instead of revising the original document and then recreating the PDF version, you can touch up the text in Acrobat using the TouchUp Text tool, one of the Advanced Editing tools.

Choose Tools > Advanced Editing > Show Advanced Editing Toolbar. The TouchUp tools are a subtoolbar of the Advanced Editing toolbar, which can be displayed separately, or select the tools from the drop-down menu (Figure 8.6).

Select the TouchUp Text tool ![icon] from the Advanced Editing toolbar or the TouchUp toolbar and click the tool on the document page within the text you want to edit. The paragraph is surrounded by a bounding box. Drag the I-beam pointer to select all or part of the paragraph, or position the I-beam within the text you want to edit (Figure 8.7). Type replacement text or add new text at the position of the I-beam pointer; click outside the highlighted area to deselect the text.

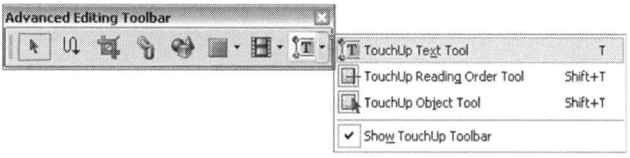

Figure 8.6 Acrobat includes three advanced TouchUp tools

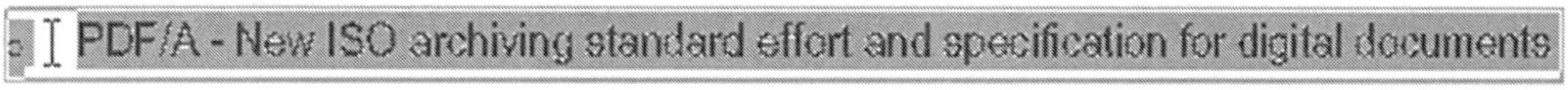

Figure 8.7 Select the text for modification using the TouchUp Text tool

The TouchUp Text tool can also be used to add new text to a page by following these steps:

1. Select the TouchUp Text tool, and then Ctrl-click the page where you want to add the text.
2. The New Text Font dialog box opens, showing the default options which are Arial text and horizontal writing mode. Select the font and writing mode (horizontal or vertical), and click OK.
3. The default text "New Text" displays on the page at the location you clicked. Select the default text, and then type the new text.
4. Click outside the new line of text to finish the process.

If the line of text becomes too wide for the page layout, you can add line breaks. After adding the text, click on the text where you want to break the line, right-click to open the shortcut menu, and choose Insert > Line Break and press Enter or Return to wrap the text to the next line. Use the same process to insert other items such as soft hyphens, nonbreaking spaces, and em dashes.

If you need to adjust paragraphs to make the layout fit properly, you can use the TouchUp Object tool, described later in the chapter.

Modifying Text Attributes

You can modify properties of existing text as well as new text added to a document in a number of common ways, such as the font and font size, fill and stroke, spacing, and baseline adjustments.

Use the TouchUp Text tool to modify text attributes by following these steps:

1. Select the text or characters you want to edit with the TouchUp Text tool.

2. Right-click the text to open the shortcut menu and choose Properties; the TouchUp Properties dialog box opens (Figure 8.8).

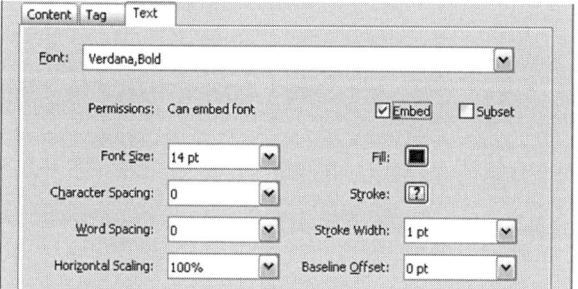

Figure 8.8 Modify attributes for selected text

3. Choose settings as required. For example, click the Font drop-down list and choose an alternate font if necessary. The fonts currently in use appear first in the list, followed by other fonts on your system.

4. As you make adjustments, the changes are automatically previewed in the selected text (Figure 8.9).

5. Click Close to dismiss the dialog box and apply the settings

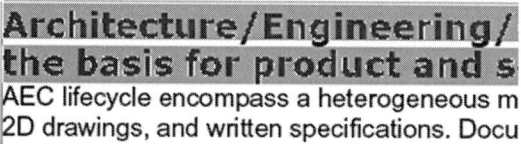

Figure 8.9 Acrobat shows a dynamic preview as you adjust attributes

TouchUp Text Results

Your source program's fonts may not convert as you expect in all cases. What you may define as bold or italic text in a source Word document may actually be a simulated appearance. Unless you have specified a font that is a named font, such as Arial Bold, touching up the text in Acrobat won't display the simulated appearance. For predictable results, make sure to embed and subset fonts before conversion if you anticipate making text changes in Acrobat.

Object TouchUps

Objects created in Acrobat, such as form fields, can be selected with their associated tool or the Select Object tool 🔲, located on the Advanced Editing toolbar. Select and manipulate content imported as part of the document, such as text or images, using the TouchUp Object tool.

The TouchUp Object tool 🔲 is located on the TouchUp subtoolbar of the Advanced Editing toolbar, and is very useful for organizing page content. If you have added extra text, for example, you can use the TouchUp Object tool to select blocks of text and drag to adjust the page's content (Figure 8.10).

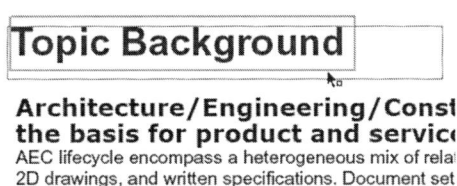

Figure 8.10 Use the tool to rearrange content on the page

Once an object is selected it can be cut, copied, and pasted. You cannot select an object on one page and drag it to another page. However, you can select and cut an object from one page and then paste it to another page. The TouchUp Object tool will select multiple types of object; to select content quickly you can click and drag a marquee with the tool that selects all the objects within the marquee.

Other Uses

Use the TouchUp Object tool to do other types of editing in your document aside from common copy/paste functions. Select an object or objects, and choose these edit options from the shortcut menu:

- Delete Clip removes objects clipping the selected object.
- Create Artifact removes the selected object from the reading order to prevent it from being identified by a screen reader or the Read Out Loud feature.
- Edit Image to edit a bitmap in Photoshop or Edit Object to edit a vector object in Illustrator – the commands change depending on the object type selected.

TouchUp Preferences

If you want to use external programs for editing images and text, specify the programs in the preferences. Choose Edit > Preferences > TouchUp. For either the Choose Image Editor or Choose Page/Object Editor options, click the button to open a browse dialog. Locate and select the program to use and click Open to close the dialog box and assign the program to the function.

Touching Up Reading Order

The final TouchUp tool is used to touch up reading order; that is, the path which a reading device will follow through a document and its pages. The TouchUp Reading Order tool is used to define content elements on a page and then order them as necessary.

To evaluate and adjust the reading order in a page, select the TouchUp Reading Order tool on the Advanced Editing toolbar's TouchUp subtoolbar and click the document page with the tool to open the TouchUp Reading Order dialog box.

The page content is shown in numbered gray blocks identifying the reading sequence (Figure 8.11).

Although an in-depth discussion on working with Reading Order and other accessibility issues is beyond the scope of this book, here are some fundamentals for working with Reading Order:

- The default reading order of a page is assigned from left to right, top to bottom. Assign a different order in the program preferences by choosing Edit > Preferences > TouchUp. Choose an alternate order from the TouchUp Reading Order drop-down menu.
- If reading order is important in your situation and the page contains many extraneous/incorrect tags, clear the structure on the TouchUp Reading Order dialog box and rebuild the page manually.
- Extra tabs, lines, and spaces added in the source document are included in the reading order. Tag the object as an artifact or remove it from the document's contents.

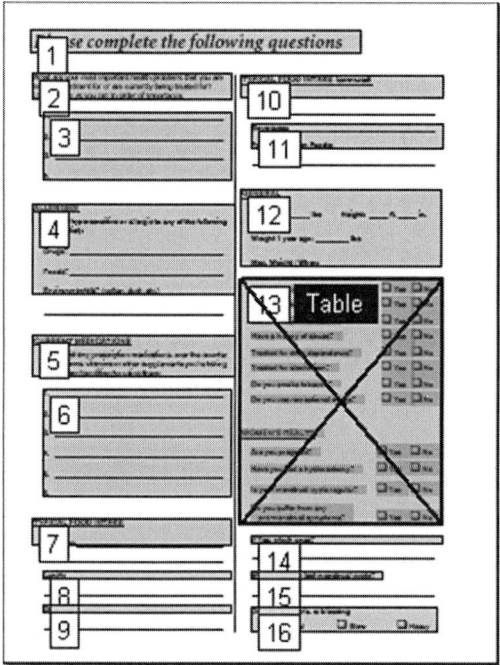

Figure 8.11 The Reading Order is defined on each page

Exporting Documents from Acrobat

Just as you can create a PDF file in a number of ways from a range of source programs and file formats, you can also export a PDF file from Acrobat in a range of different file formats. You can, for example, create a PDF file from an Excel spreadsheet using the PDFMaker installed in Excel, and then export the file from Acrobat as a Web page in HTML format.

To save a PDF file in another file format, follow these steps:

1. Choose File > Save As to open the Save As dialog box (Figure 8.12).

2. Click the Save as Type drop-down arrow to open the list and choose a file format.

3. Click Settings to open a Settings dialog box with options specific to the chosen file format.

4. Modify settings as necessary and click OK to return to the Save As dialog.

5. Click Save. Acrobat exports the file in the chosen format and the Save As dialog box closes.

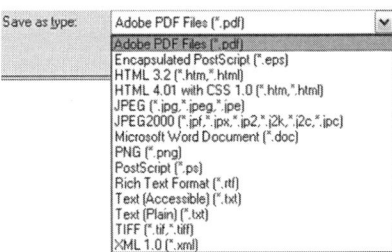

Figure 8.12 Export files in a wide range of formats

Round-trip Editing

Use the TouchUp Object tool and a designated source program, such as Photoshop, to make image changes that are returned to the PDF document by following these steps:
1. Select the TouchUp Object tool and then select the image or images you want to edit.
2. Right-click and choose Edit Image from the shortcut menu to open the designated editing program and displays the image or images.
3. Make the edits; flatten layers if you have made changes to the layer structure.
4. Choose File > Save. The image is saved, closed, and replaced in Acrobat.
    There are some other issues to consider when using Photoshop as an image editor. The image connection only exists as long as the object is selected in Acrobat – do not deselect the object in Acrobat or you have to start the procedure over. Changing image modes may not be saved automatically; a Save As dialog box opens to save the image as a separate Photoshop PDF file. If the image looks distorted, then check the pixel aspect ratio, as Acrobat forces Photoshop to use pixel aspect ratio correction for previewing. Check your Photoshop configuration if you see a checkerboard instead of the image in Photoshop.

Exporting Files for Use in a Document-processing Application

Acrobat allows for two export formats that you can use in Word or other document-processing programs. When the document is open in Acrobat, choose > File Save As to open the Save As dialog box. For export to a document-processing application, choose either rich text format (RTF), or Word document (DOC) formats. Click the Settings button on the Save As dialog box to open a set of options(Figure 8.13).

Configure the export settings depending on your intended use of the file:

- Include Comments is selected by default; deselect it if you do not need to use comments in the exported document.
- Maintain the layout structures in the PDF file if it contains specific features like custom margins, columns, and so on. The options for DOC export are slightly different than those for RTF export.
- Deselect the image export options if you do not need them in the exported file. If you want to export images and your file contains both grayscale and color images, choose Determine Automatically from the Use Colorspace drop-down list.
- The option for generating tags is a default selection – the tags are used in the conversion process and then discarded. Leave the option selected.

Exporting Text or Web Formats

You can export the content and images from a PDF file from Acrobat in one of two HTML formats, XML, or accessible or plain text. Neither Acrobat 6 nor Acrobat 7 support XHTML exports, or any CSS version higher than 1.0.

Choose File > Save As and select a file format option from the Save As drop-down list. Click Settings to open the Settings dialog box specific to the selected format type; if you choose accessible text, there are not any settings to modify. Figure 8.14 shows the settings for an HTML 4.01 with CSS file export. Acrobat uses the default mapping table for the file's content, but you can choose an alternate method from the Encoding drop-down list.

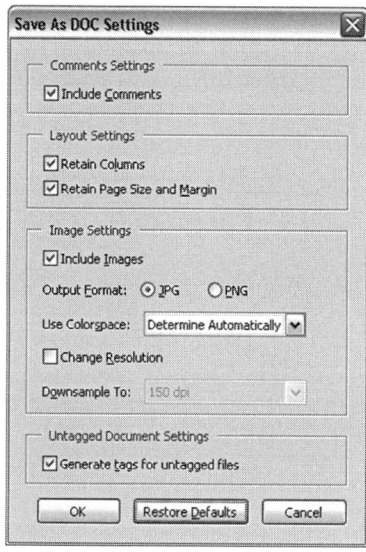

Figure 8.13 Choose export options for document-processing application use

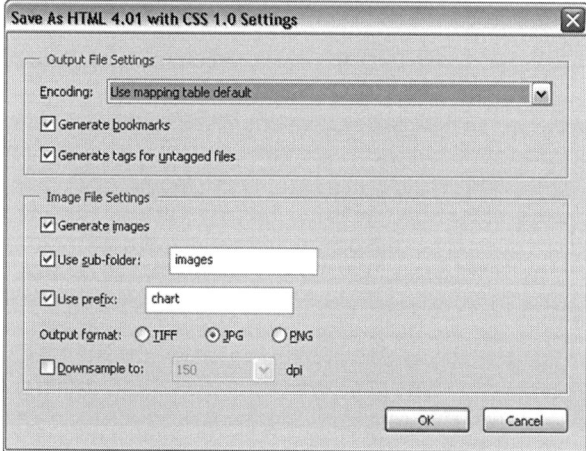

Figure 8.14 Customize settings for Web page exports

In all cases, tags are generated automatically. Bookmarks are converted to links and placed at the start of the file. When exporting images, Acrobat creates a subfolder named "images" to store the exported image files. Images exported from a PDF file in this manner are numbered and use a prefix if specified in the Settings dialog box.

Exporting a PDF File as Images

Instead of exporting a PDF file as a combination of text and images, you can export it as a single image per page, which you might choose to use the content in another project or to create thumbnails to use in other applications, such as visual links. If you want to protect the content of the document, exporting it as an image with an attached security policy prevents manipulation of the content of the file.

To save a PDF file as images, choose File > Save As and select an image export option from the Save As drop-down list. Click Settings to open the Settings dialog box and choose the options according to your intended use. Acrobat will convert each page to an image the same size as the document page.

Exporting Images from a PDF File

Sometimes you may not have the source files available, but have images in a PDF file that you'd like to reuse in another application. You can export the images from a PDF file – without the text – and specify the minimum size for image extraction. The extraction process supports JPEG, PNG, TIFF, and JPEG2000 image formats. JPEG and JPEG2000 images having a specified compression and resolution aren't affected by the settings you choose in the Settings dialog box.

Follow these steps to extract images:

1. Choose Advanced > Export All Images to open the Export All Images As dialog box. Locate and select the folder you want to use to store the images.

2. Choose an image format from the drop-down list at the bottom of the dialog.

3. Click Settings to open the Export All Images As dialog box (Figure 8.15). Specify export settings for the images; click the Extraction drop-down arrow and choose an image size for exclusion. The default size is 1.00 inch, meaning images equal to or less than 1 inch are not exported. Excluding images based on size is very useful if you are extracting images from a file containing a repeating small image, such as a company logo.

4. Click OK to close the Settings dialog box, and then click Save to export the images to the specified folder. The images are saved using the file's name and an incremental number.

Attaching Source Files to a PDF File

Source and ancillary information files can be attached to a PDF document, and viewed in Acrobat 6 or 7, as well as Adobe Reader. Attaching additional files is a useful way of compiling files for a project, as well as storing content in a single, easy-to-access location.

The simplest method for attaching a source file to a PDF file is by modifying the PDFMaker settings before converting the file to PDF. In a program using a PDFMaker, choose Adobe PDF > Change Conversion Settings to open the Acrobat PDFMaker dialog box and click Attach source file to Adobe PDF in the Application Settings area of the Settings tab. Click OK to close the dialog box, and proceed with the conversion.

Within Acrobat, you can attach a file to the document using the Attach tool, or via the Attachments pane. Click the Attach tool 📎 on the File toolbar to open the Add Attachment dialog box. Locate and select the file to attach and click Open. The dialog box closes, and the file is attached to the PDF file.

Export Preferences

If you pay attention to how you ordinarily work with files, you may find that you are performing many of the same types of file export, using the same settings customizations. Rather than repeatedly modifying export settings for a particular file format, change the Acrobat program preferences by following these steps:

1. Choose Edit > Preferences and click Convert from PDF in the left pane of the dialog box.
2. Select the format you want to modify from the list in the right pane to view its settings.
3. Click Edit Settings to open the settings dialog boxes used to configure export of an individual file.
4. Adjust the settings as required and click OK to close the settings dialog box.
5. Click OK to close the Preferences dialog box.

The next time you export a file of that type, the modified settings are applied, saving you processing time.

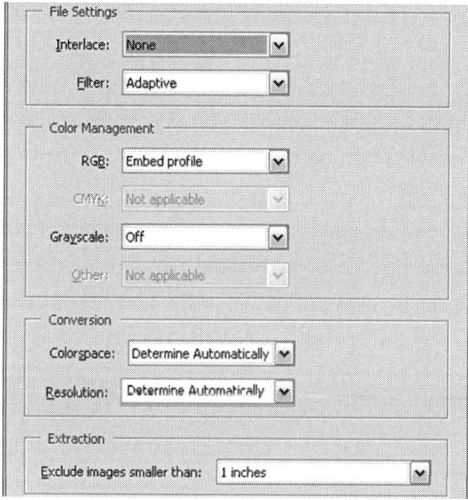

Figure 8.15 Choose export settings for images saved from a PDF file

Attaching a File as a Comment

You can indicate an attached file corresponding to specific information in your document using a file comment rather than attaching a file to the overall PDF document. For example, attach a file to a page at the location where a reference is cited, the reader clicks the comment icon to open the source file. Files attached as comments are listed in both the Attachments and Comments panes. You cannot attach an open PDF file to another PDF document, but you can attach a file of any format.

Follow these steps to attach a file as a comment, and customize its appearance:

1. Click the Attach icon's drop-down arrow on the File toolbar and click the Attach a File as a Comment tool 📎 .

2. Click the document with the tool at the location where you want to attach the file to display the Add Attachment dialog box. Locate and select the file to attach and click Select to close the dialog box, which is replaced by the File Attachment Properties dialog box.

3. In the File Attachment Properties dialog box choose an alternate icon, color, or opacity in the Apperance tab of the dialog box if desired.

4. On the General tab of the File Attachment Properties dialog box, modify content describing the attachment, such as the author's name and a description.

5. Click Close to close the dialog box when the customization is complete.

The attachment icon displays at the page location clicked with the tool. Move your mouse over the comment icon to display the name and description in a tooltip (Figure 8.16).

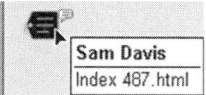

Figure 8.16 View the comment author's name and a description of the attached file in a tooltip

Managing Attached Files

A document containing attachments, regardless of whether they are attached to the document or embedded as a comment, displays a paperclip icon ✎ at the lower left of the program window.

Display the Attachments tab horizontally below the Document pane by choosing View > Navigation Tabs > Attachments or select the Attachments tab in the Navigation tabs at the left of the program window. The Attachments pane shows basic information about the attachments, such as their name, description, size, and modification date (Figure 8.17).

The Location in document column shows where the attachment originated. Files added during a PDFMaker conversion process and those added as attachments in Acrobat list the location as Attachments tab. Any files that are added using the Attach File as Comment tool list the attachment's page number.

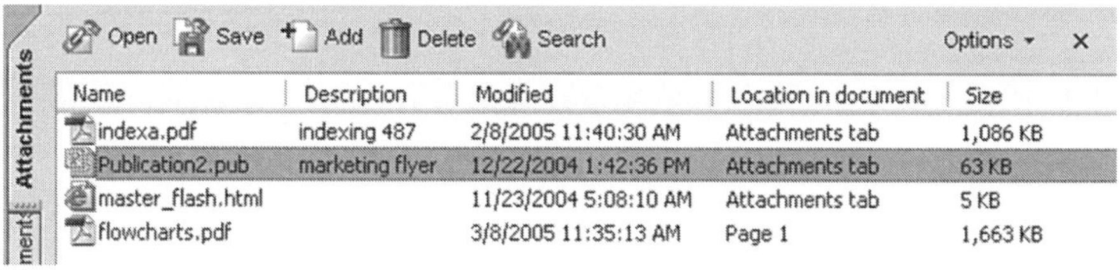

Name	Description	Modified	Location in document	Size
indexa.pdf	indexing 487	2/8/2005 11:40:30 AM	Attachments tab	1,086 KB
Publication2.pub	marketing flyer	12/22/2004 1:42:36 PM	Attachments tab	63 KB
master_flash.html		11/23/2004 5:08:10 AM	Attachments tab	5 KB
flowcharts.pdf		3/8/2005 11:35:13 AM	Page 1	1,663 KB

Open   Save   Add   Delete   Search   Options ▾   ✕

Figure 8.17 Attached files are listed in the Attachments pane.

Attachment Actions

Click the attached file's name and choose commands from the menu for managing the attachment, right-click to display a shortcut menu, or use the corresponding icon on the Attachment pane's toolbar.

You can apply several commands to a selected attachment including:

• Click Open on the Attachment pane's toolbar or double-click an attachment in the list to open it. PDF files open in Acrobat, while files of other formats open in their associated source programs. When you try to open a file in another format, Acrobat displays a warning dialog and a list of actions (Figure 8.18).

- Save attachments independently of the parent PDF document. Click Save on the Attachment pane's toolbar or from the Options or shortcut menus to open the Save Attachments dialog box. Choose a name and folder location and click Save to close the dialog box and save the attachment.
- Click Add on the Attachments pane's toolbar, or click the Add Attachment icon on the File toolbar to open an Add Attachment dialog box. Locate and select the file, and click Attach to close the dialog box and attach the file.
- Delete attachments by selecting them in the Attachments pane and clicking Delete on the Attachment pane's toolbar or press the Delete key.
- Add a descriptive label by choosing Edit Description from the shortcut menu. A text entry dialog box opens, type the text, and click OK to close the dialog box and add the description. Several attachments in Figure 8.17 include descriptions.

Viewing Attachments

You can configure a document to automatically display the Attachments pane when the PDF document is opened. Choose File > Document Properties > Initial View. Click the Show drop-down arrow and choose Attachments Panel and Page in the Document Options area of the dialog box and click OK (Figure 8.19)

If you have the Attachments pane open, click the Options menu and choose Show attachments by default; saving and reopening the file shows you the selected opening view.

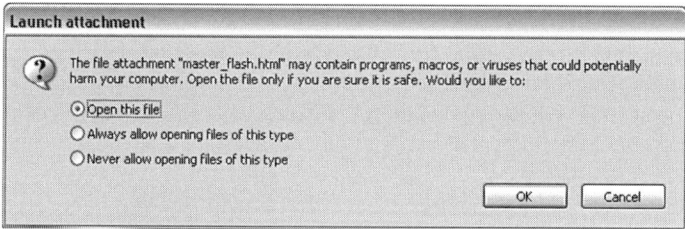

Figure 8.18 Choose options to apply in the current and future work sessions

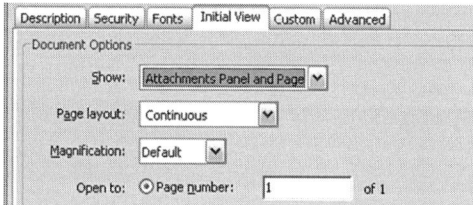

Figure 8.19 Set the initial view to display attachments automatically

Using Attachments in Earlier Versions of Acrobat

If a user working with Acrobat versions 5 or 6 opens a PDF document containing attachments, an information dialog box describing the attachments displays before the file opens. In Acrobat 5 or 6, choose Document > File Attachments to open the File Attachments dialog box which lists the attachments according to their locations. Document file attachments are listed first followed by those attached as file comments listed by page number.

Incorporating Attachments into a Workflow

Using attached files is convenient for storage and distribution. Consider incorporating attachments into your workflow if you need to:
- Control content attached to the source file, as you can readily see the attachments listed in the Attachments pane in Acrobat.
- Maintain file integrity, as moving files always moves the attachments with the parent PDF file.
- Ensure security, as attached files can be protected during emailing, as described in Chapter 10.
- Search content, as attached files can be searched in Acrobat.
- Supply ancillary information in its native formats, as attachments are stored with the PDF file, not converted to PDF format.

Summary

In this chapter we discovered a range of tools and processes you can use to manipulate the content and indeed, the type of file you are working with. You saw how to reuse text, images, and tables from a PDF file. You saw how the TouchUp tools can be used to modify existing content in a document or add additional objects.

You learned how a PDF file can be exported in a wide range of formats, from a Web page to an image or a series of images, and how to export just the images from a file. Finally, you learned how to work with files – both in PDF and other formats – that can be attached to a PDF file.

Exercises

1. Using the methods outlined in the chapter, experiment with the content of an existing PDF file by selecting, exporting, and adding text to the file.

2. Try exporting a table from a PDF file that is tagged, and also one that is not tagged. Can you see the difference in the use of the material?

3. Save a sample PDF file in various export formats. Can you see how you could integrate some of these features into your workflow?

4. Extract the images from a document, and experiment with the exclusion size for the images.

5. Try attaching files to a PDF file, both before conversion working with the PDFMaker in a source program, and also within the Attachments pane in Acrobat.

Project

Use the files in the **ch08_project** folder, which include three source files and one PDF file.

Task 1: Reusing content

In this task you see how any copied content can be used as a separate document.

1. Click the Select tool on the Basic toolbar, and then click and drag (or use another selection method) to select about one-half the text from the first page.
2. Right-click the selection and choose Copy to Clipboard from the shortcut menu.

3.   Open Word, and a blank document. Choose Edit > Paste and paste the text.

Task 2: Extracting a new document

In this task, you extract pages from the **Soccer Park.PDF** document and save them in another file format.

1.   Choose Document > Extract Pages. In the ensuing dialog box, enter pages 1 to 6 as the start and end pages for extraction.
2.   Click OK to extract the pages to a separate PDF file.
3.   Choose File > Save As to open the Save As dialog box, and choose Microsoft Word Document (.doc) from the File type drop-down list.
4.   Click Save to save the file in the selected format. When you open the file in Word you see it is identical in layout to the PDF file.

Task 3: Extracting a table

In this task you select the table and open it in Excel, separate from the rest of the document.

1.   Click the Select tool on the Basic toolbar, and then select the table on page 13 of the **Soccer Park.PDF** file.
2.   Right-click the selection to open the shortcut menu, which includes three table choices (Copy as Table, Save as Table, and Open Table in Spreadsheet).
3.   Choose Open Table in Spreadsheet to place the contents of the table into an Excel spreadsheet.

You see that the figures are laid out in the spreadsheet correctly, but notice that the dollar signs are added in separate columns, and the columns used to contain headings for other columns of figures.Clearly the table was built by someone who wasn't a skilled spreadsheet creator!

Task 4: Reutilizing a photograph

Page 16 of the Soccer Park.PDF file is a photograph. In this task you reuse the photograph in two different ways by copying to the Clipboard and by using a Snapshot.

First follow these steps to use the Clipboard method:

1.   Click the Select tool on the Basic toolbar, and then click the image on page 16 of the **Soccer Park.PDF** file.
2.   The image is highlighted; a Copy Image to Clipboard icon appears at an upper edge of the image.
3.   Click the Copy Image to Clipboard icon to preserve the image.
4.   Paste the image from the clipboard into any file that can accept a JPEG file.

Rather than using the whole image, use only a select portion working with the Snapshot tool:

1.   Zoom in to the image at a desired magnification. The magnification displayed is used by the tool.
2.   Click the Snapshot tool on the Basic toolbar to select it.
3.   Drag a marquee around the portion of the photograph you want to reuse, and release the mouse to copy the selection to the clipboard.
4.   Paste the image into any file that can accept a JPEG image.

You can also use the captured image as a stamp by choosing Tools > Commenting > Stamps > Paste Clipboard Image as Stamp Tool.

Task 5: Adding attachments to the PDF file

The three original files used to create the **Soccer Park.PDF** file are in the **ch08_project** folder. Attach the files to the PDF document for reference.

1. Select View > Navigation Tabs > Attachments to open the Attachments pane.
2. Click Add to open the Add Attachment dialog box. Locate and select the first file for attachment, **03300 Cast-In-Place Concrete.doc**.
3. Click Open to close the dialog box and attach the document to the PDF file.
4. Repeat with the spreadsheet and image files.

9

Adding Navigation to the Package

How often have you downloaded a PDF from the Web only to have it be just like a paper book? All you could do is turn the pages – the document's author left out many great features of Acrobat, and in so doing sold the work short.

Acrobat contains navigation features that let you design files that work similarly to Web pages. The navigation features give you extra control of the document, and, as a result, more power in conveying your information. The most commonly used navigational structures in a PDF document are links and bookmarks.

Links work in much the same way as those seen on a Web page, but can also be used with a number of alternative actions. Bookmarks are a type of navigational structure that use a navigation panel to link content based on a structural hierarchy created manually, from styles or headings in a source document, or derived from the document's structure. Like links, bookmarks can be used with a variety of actions beyond simple navigational hyperlinks.

A PDF document, with all its features, more closely resembles a Web site than an ordinary printed document. You expect a Web site to have a navigational structure; you can design much the same structure in a PDF document as well.

In this Chapter

In this chapter you see how to add navigation to allow your reader to move through your document as you learn how to:

- Design a navigation strategy for a project
- Build and configure bookmarks
- Create bookmarks using a tagged file
- Set links and add actions
- Use buttons to control actions and navigation in a document
- Create duplicates of buttons throughout a project.

Strategy

Before randomly adding navigation features such as links or bookmarks to a document, plan a strategy for your document or document collection. Your plan should consider several factors, including the content of the document, how you project the user will navigate through the file, and the storage plans for the document or documents. Should the package be a single file or multiple files?

The answer depends on how much control you will have over the published package. Is there a chance someone will accidentally move a file or folder? If a file or folder is moved independently of the rest of the package, then links and bookmarks can be broken, resulting in errors.

As you become familiar with Acrobat you will, of course, become more adept at planning a document's navigation structure. For the beginner, a single file is probably the simplest approach, but keep in mind that single files compiling information from an engineering/construction project can easily number 4,000 or more pages.

Tom's first major PDF project was for the US Air Force: a single PDF file containing 3,383 pages. The single file contained the entire project, and included CAD files from both MicroStation and AutoCAD, source material from seven other programs, and scanned documents.

You may well ask yourself, "Doesn't it take a lot of time to build a big document like this?" The answer is "yes" and "no." If the clients intend to manage the project electronically, compiling all source material into PDF format saves a tremendous amount of time over the paper route. In Tom's example, the Air Force was not yet ready for electronic project management, so everything was done in both paper and PDF format. To add navigation, each inserted document had its own bookmark, identifying its topic. All the reader need do is click the appropriate bookmark to display the document in Acrobat. The user is not going to page through a document of this size – navigation becomes the key.

Acrobat 6 versus Acrobat 7

In Acrobat 6, the link is set to a page number, so adding or deleting pages changes the link's location. To ensure consistency in a link, you have to set named destinations. Advancements in Acrobat 7 let you set links to a page view that is very similar to using a named destination, but many times easier.

Bookmarks

In Acrobat, bookmarks are the digital model for the pieces of paper you stick between pages to mark important information. Unlike paper, a set of bookmarks in Acrobat is much neater and quite orderly. Bookmarks are created in a variety of ways, as you will see in this chapter.

The Bookmarks pane shows a listing of linked bookmarks. The [+] in front of a bookmark means there are nested bookmarks within the named bookmark that you see on expansion. A bookmark hierarchy that is expanded shows a [–] to the left of the bookmark's name (Figure 9.1). The structure of the bookmark hierarchy can be adjusted to change the hierarchical relationships between the bookmark levels, as you learn later in the chapter.

About the Example

The project used for these screen shots is a 3,875-page Electronic Owner's Manual for a high school. PDF manuals can save a tremendous amount of time in education maintenance.

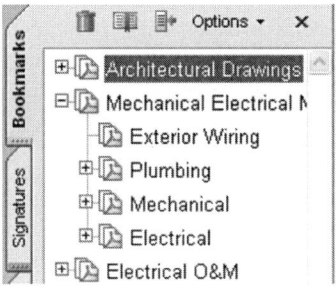

Figure 9.1 Bookmarks are displayed and nested in the Bookmarks pane

Specifying Opening View

By default a PDF file opens with only the Document pane displayed. Many people do not know the bookmarks navigation even exists, or are not aware how to display the pane. Show your users the navigation provided by configuring the way the document opens. The document opening view you specify is not in effect until you save, close, and then reopen the document again.

To specify that a document should display the Bookmarks pane when it opens, follow these steps:

1. Choose File > Document Properties to open the Document Properties dialog box and click the Initial View tab (Figure 9.2).

2. Click the Show drop-down arrow and choose Bookmarks.

3. Click the Magnification drop-down arrow and choose a zoom factor for the document.

4. Click OK to close the dialog box and save the file.

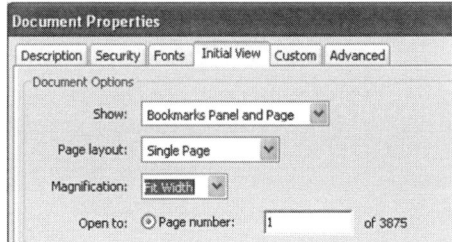

Figure 9.2 Set the view displayed when the document opens

After saving and reopening the file, the Bookmarks are shown with the Document pane displayed according to the specified magnification.

Specifying a View for Other Pages

Sometimes the creator does not want the bookmarks to open when the document opens to the initial page. You can set the bookmarks to open with another page by using an action. Although bookmarks default to showing the first page of a document at a specific magnification, you can add a great number of different actions to the bookmark that are triggered in response to activities such as a user's mouse click.

In this example, viewing the second page of the document triggers the execution of a menu item that displays the Bookmarks pane.

Follow these steps to specify an alternate starting page for bookmarks:

1. Select the Pages tab to display the Pages pane. Select the Page 2 thumbnail, and right-click to display the shortcut menu.

2. Click Properties to open the Page Properties dialog box, and select the Actions tab.

3. Click the Select Trigger drop-down arrow and choose Page Open from the list. When the selected page (page 2) is opened in the program window, the action is triggered.

4. Click the Select Action drop-down arrow and choose Execute a menu item.

5. Click Add to open the Menu Item Selection dialog box, and choose View > Navigation Tab > Bookmarks (Figure 9.3).

6. Click OK to close the dialog box and add the action to the Page Properties dialog box.

7. Click Close to dismiss the dialog box and set the action.

8. To test the action, save the file, close it and reopen it. When you navigate to page 2 in the document, the Bookmarks pane will display in the program window.

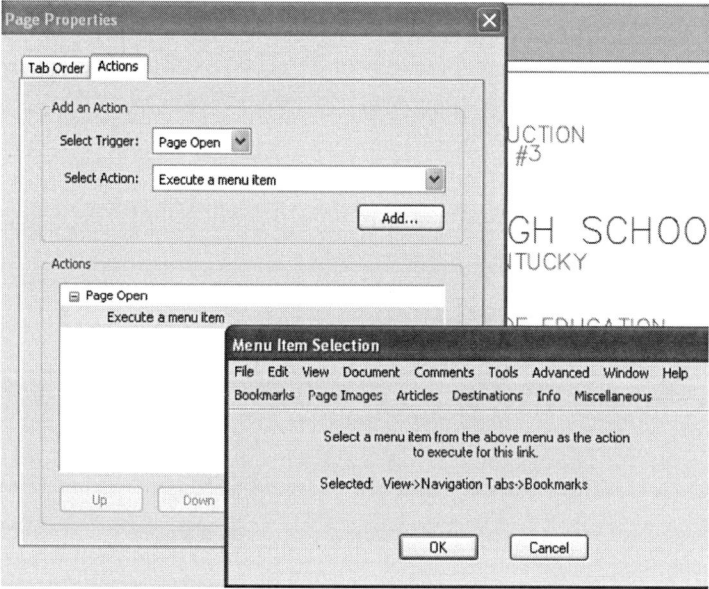

Figure 9.3 Use page actions to customize bookmark viewing

Creating Bookmarks

Bookmarks can be created in a number of ways; creating bookmarks using PDFMaker settings was discussed in Chapter 4.

Manually creating bookmarks is simple when the document contains text. Follow these steps to create a bookmark using selected text:

1. Using the navigation controls, scrollbars, and Hand tool, position the page as you want it to display when the bookmark is clicked.

2. Click the Select tool 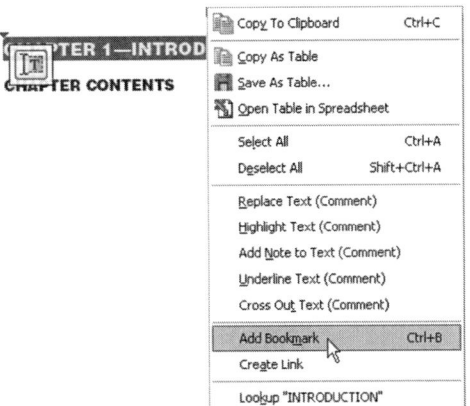 on the Basic toolbar, and then drag to select the text to use for the bookmark.

3. Right-click to display the shortcut menu, and choose Add Bookmark (Figure 9.4).

4. The dialog box closes, and a new bookmark using the selected text is added to the Bookmarks pane.

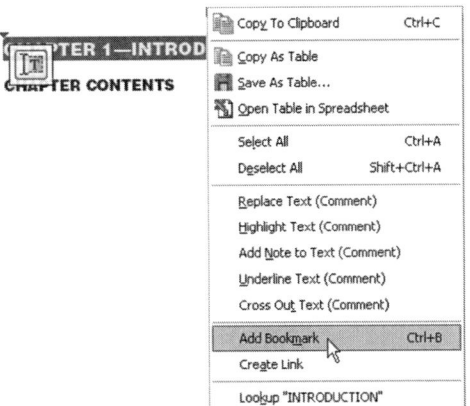

Figure 9.4 Use highlighted text to create a bookmark

Note: Click the bookmark prior to the location where the new bookmark will be added. If there is no selected bookmark, a new bookmark is added to the end of the list.

To add an untitled bookmark and change the text at another time, follow these steps:

1. Position the document in the Document pane, and select a bookmark in the Bookmarks pane if you want the new bookmark placed in a specific location (Figure 9.5).

2. Click the New Bookmark icon 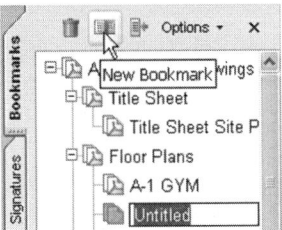 on the Bookmark pane to add an Untitled bookmark.

3. The default text is active; type a name for the bookmark.

Figure 9.5 Add an Untitled bookmark and then type a text label

Adjusting Bookmarks

You can customize bookmarks added to a document in a number of ways. You can adjust the order of the bookmarks in the list, or modify the hierarchy by demoting or promoting bookmarks. You can also customize the font style and color of bookmarks for emphasis.

Reordering and Nesting Bookmarks

When you have finished adding bookmarks to a file, check the order. Especially in situations where you have added numerous bookmarks manually, you may find they are out of page order. In some cases the page order isn't relevant; in most cases, it is most logical for the user to list bookmarks corresponding to the order in which they appear in the document. Drag a bookmark up or down in the list to reorder it.

In the majority of documents, all headings do not have equal weight, and the same principle applies to a corresponding set of bookmarks in the PDF version of the file. If you have converted a file using a PDFMaker and specified that headings or styles are used for bookmarks, you already have a bookmark hierarchy.

In projects where you are adding bookmarks manually, you can emulate the same nesting arrangement for bookmarks. To nest a bookmark, select it in the list and drag it to the chosen location; if you drag the indicator arrow and line below an existing bookmark, it will be nested below the bookmark when you release the mouse (Figure 9.6). To promote a bookmark, drag it to the left as you move it within the hierarchy.

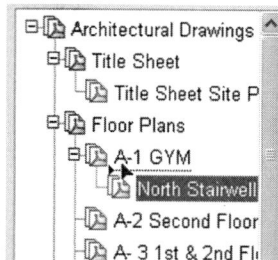

Figure 9.6 An indicator shows the proposed location for a nested bookmark

In previous versions of Acrobat, bookmark organization used similar processes; however, the techniques used in Acrobat 7 are considerably more intuitive.

Color Coding Bookmarks

Acrobat allows color coding of bookmarks to add an additional level of information and place emphasis on higher levels of bookmarks. An AEC example might be changing the color of the bookmarks of all design sheets that have a fire code issue to red.

Follow these steps to color-code bookmarks in a document:

1. Select the bookmark in the Bookmarks pane.

2. Right-click the bookmark and choose Properties from the shortcut menu to open the Bookmark Properties dialog box (Figure 9.7).

3. In the Appearance tab, select a Style from the drop-down arrow, which is Plain by default. For emphasis, you can choose Bold, Italic, or Bold Italic text options.

4. Click the Color swatch to open a color palette and choose an alternate color. If you want a custom color, click Other Color to open a larger Color Picker.

5. Click OK to close the Bookmark Properties dialog box and set the bookmark's appearance.

Note: If you have several bookmarks in the file that you would like to apply similar formatting to, Ctrl-click the bookmarks in the Bookmarks pane before opening the Bookmark Properties dialog box.

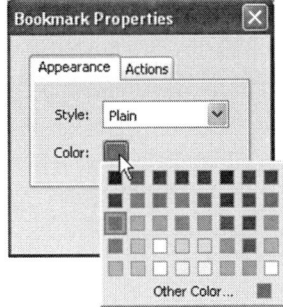

Figure 9.7 Customize the appearance of some bookmarks for emphasis

Changing Properties

If you want to use a custom appearance for all the bookmarks, perhaps using a custom color to match color used in the document, you can define the details when you start. Add the first bookmark and then configure it in the Bookmarks Properties. Right-click the bookmark and choose Use Current Appearance as New Default.

Rather than opening dialog boxes to make changes to document elements, you can use Acrobat's Properties bar. Right-click the toolbar well at the top of the program window to show the list of menus and select Properties. The toolbar's name and contents change according to the object selected. For example, in Figure 9.8, the upper image shows the toolbar when a bookmark is active in the document; the lower image shows the toolbar when a link is active in the document.

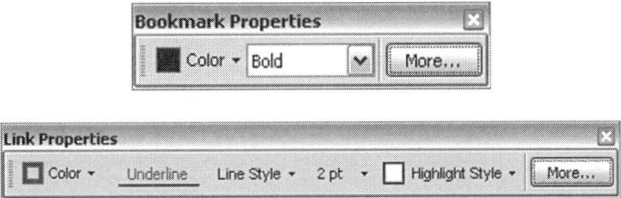

Figure 9.8 The Properties toolbar varies according to the content selected

Using Tagged and Web Bookmarks

The Bookmarks pane displays one of three icon before a bookmarks' name, which are functionally different, although you can modify them in the same way. The options include the default bookmark icon 🔖, a tagged bookmark icon 🔖, or a Web-page generated bookmark 🔖.

Adding and manipulating basic bookmarks has no effect on the document structure, while tagged bookmarks can be used for modifying a document's content as well as navigation. Documents can be exported from Microsoft Word XP or Adobe InDesign CS as tagged documents without bookmarks assigned to styles or headings.

Follow these steps to build bookmarks from a tagged document:

1. From the Bookmarks pane, choose Options > New Bookmarks from Structure to open the Structure Elements dialog box.

2. Select the tags you want to convert to bookmarks. Ctrl-click to select non-contiguous tags. Choose tags according to the levels of headings you want in your Bookmarks list (Figure 9.9).

3. Click OK to close the dialog box. In the Bookmarks pane, a basic parent tag named Untitled is added, and the new tags based on the document structure are nested within it.

4. Expand the bookmarks list; you see all bookmarks generated from tags show the tagged bookmark icon and are named according to the tag's content (Figure 9.10).

5. Modify the bookmarks' appearance and view as desired.

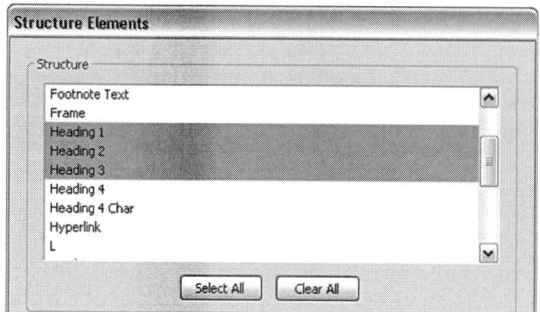

Figure 9.9 Select the elements in the document structure to convert to bookmarks

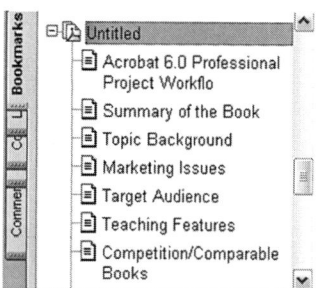

Figure 9.10 The list of tags is displayed in the Bookmarks pane

Modifying Content Using Tagged Bookmarks

You can modify the content of a document using tagged bookmarks by selecting the bookmark and choosing a command from the Options or shortcut menus. The content of your document can be modified using tagged bookmarks. Select a bookmark (or bookmarks) in the Bookmarks pane and right-click to open the shortcut menu, or click the Options button to display the pane's menu. You can click Print Page(s) to print the pages containing the selected tag(s), click Delete Page(s) to delete the pages containing the selected tag(s) or click Extract Page(s) to extract the information from the pages containing the selected tag(s) and create a new PDF document.

Web Page Bookmarks

A converted Web page automatically generates tags that can be configured, renamed, and modified, as with other types of bookmark. You can also open a link in a Web browser or append Web pages from the converted Web page's site.

Using Links for Navigation

What would the world be without hyperlinks? Acrobat offers the ability to easily create links in any document, which can link to a view or file, or initiate one of many actions.

To add a link to a document, follow these steps:

1. Choose Tools > Advanced Editing > Show Advanced Editing Toolbar to display the tools.

2. Select the Link tool ⬚ on the Advanced Editing toolbar.

3. Drag an area for the link on the document page with the tool; release the mouse to display the Create Link dialog box (Figure 9.11).

4. Choose whether the link is visible or invisible. For a visible link, select characteristics for the link's frame, such as a dashed or solid box around the text, or underline the text.

5. Choose an action from the radio buttons on the dialog box and click Next.

6. The subsequent dialog box depends on the chosen action. Choose the appropriate settings and then return to the Link Properties dialog box.

7. Click Close on the Link Properties dialog.

8. Click the Hand tool on the Basic toolbar to deselect the Link tool and test the link.

To make changes to an existing link, click it with either the Link tool or the Select Object tool to open the Link Properties dialog box again. The radio buttons available on the General tab when a new link is constructed are not shown. Instead, the dialog box displays the Actions tab.

> **Tip:** If the document has text and you want to simulate a Web page link, change the text color using the TouchUp Text Tool first, and then make the Link box invisible.

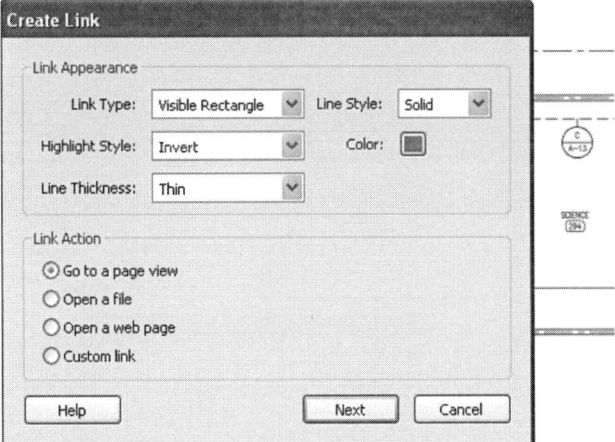

Figure 9.11 Quickly choose common settings for the link's appearance and actions

Common Link Actions

Any number of actions can be added to links, bookmarks, or buttons, but ensure the order is logical. Play a sound or movie in one document before opening another, for example. Add the actions in any order, and then click Up or Down on the dialog box to reorder the list of actions. To edit actions, click the action itself in the Link Properties Actions tab, not the details of the action, such as the file name or menu command.

The Link Properties dialog box offers lists several common actions on the General tab of the dialog box.These include:

- **Go to a page view.** To use this command, follow the instructions in the Create Go to View dialog box that displays when you select the command (Figure 9.12). Once you have set the magnification, page, and location for the link, click Set Link to close the dialog box and finish the link. In Acrobat 6, links were attached to page numbers, which was a problem if pagination was later changed; the earlier versions also required that you specify the location before starting the linking process.
- **Open a file.** If you choose this command, a file is opened when your user clicks the link. To set the link, click the Open a file radio button, which displays a browse dialog box. Locate and select the file you want to use, then specify a window open preference in the next dialog box. Click OK to close the dialog box and finish the link. Unless the link opens a PDF file, be careful when using this action, as your viewers must have the program that can open the linked file.
- **Open a Web page.** When you click this radio button, the Edit URL dialog box opens. Type a Web address in the text field and click OK to close the dialog box and finish the link. The action supports several protocols, including HTTP, FTP, and mailto. Use this command carefully – mailto tags are fairly constant, but Web pages may change on a daily basis, which makes them unsuitable for archival storage. Instead, capture the Web page and add it to the document.
- **Custom.** Click the Custom radio button to add the Actions tab to the Link Properties dialog. Click the Actions tab, and then select an action from the Select Action drop-down list (Figure 9.13) and click Add to open a corresponding dialog box. Proceed through the dialog boxes as required; read more on custom settings in the following section. Links have only one state, which means the action occurs when the link is clicked.

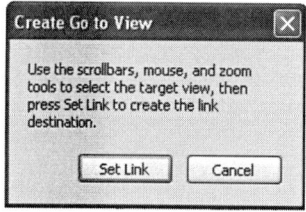

Figure 9.12 Specify the appearance of the page view as described in the dialog box

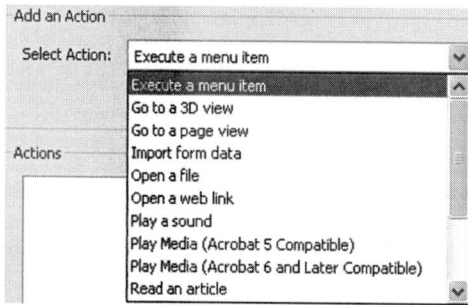

Figure 9.13 Choose an action from this drop-down list.

Tip: You can assign a bookmark to point to an image in a document. Select the image, then right-click and choose
Add Bookmark from the shortcut menu.

Other Link Actions

As shown in the previous section, choosing the Custom link radio button on the Link Properties dialog box attaches the Actions tab. The Actions tab contains the common actions that are available as radio button choices on the Link Properties dialog box. There are numerous other actions available, which are included in Table 9.1.

Table 9.1 Additional actions available for links

Selected Action	Program Activity
Execute A Menu Item	Executes an Acrobat program command chosen from listed menu selections.
Go to a 3D view	A new feature exclusive to Acrobat 7 for manipulating content in 3D space; 3D views are discussed in Chapter 6.
Import Form Data	Stored data are imported from an external file to populate fields on an active form.
Play A Sound	A selected sound file is embedded in the PDF file in a cross-platform format.
Play Media (Acrobat 5 Compatible)	Requires a linked QuickTime or AVI movie that uses Acrobat 5-compatible settings in the file; the action plays the media when the link is clicked.
Play Media (Acrobat 6 Compatible)	Link or embed a movie using Acrobat 6-compatible settings in the file; then use this action to play the media when the link is clicked (Acrobat 6 embeds only).
Read An Article	Display document content as an article thread in a defined document (rarely used in AEC).
Reset A Form	Delete content added to specified form fields.
Run A JavaScript	Runs a selected JavaScript. The most commonly used programming language for PDF forms and documents is JavaScript.
Set Layer Visibility	Define layer settings in the PDF file, and then use this action to control the active layer settings.
Show/Hide A Field	Use this action to toggle field visibility; the action is commonly used with form fields to show or hide content.
Submit A Form	Sends form data to a specified URL.

Navigating with Buttons

Using a button is an often overlooked navigation option in Acrobat. Buttons are created using the Button tool ▣ found on the Forms toolbar, a subtoolbar of the Advanced Editing toolbar (Figure 9.14).

A button has the same action capabilities as a link, but using buttons is simpler than links when you want to use the same navigation option on multiple pages.

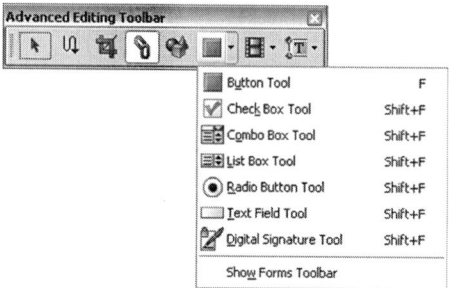

Figure 9.14 The Button tool is located on a subtoolbar

Here is an example: suppose you are working with a manufacturer's equipment manual that has a linked Table of Contents, and no bookmarks. You could easily add a button to every page of the manual that you can use to display the Table of Contents.

Follow these steps to add and configure the first button for the manual:

1. Click the Button tool on the Forms subtoolbar and draw a button in one corner of the document page. In Figure 9.15, the button is drawn at the upper right corner of the page.

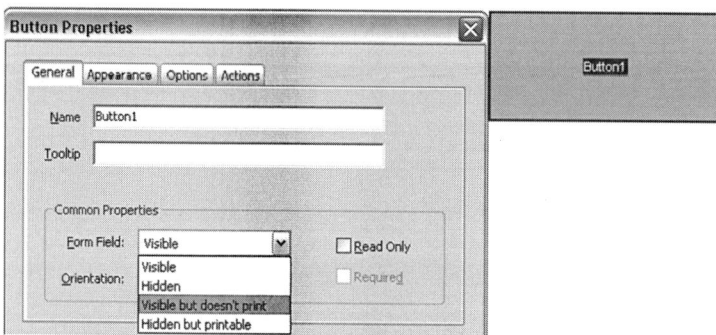

Figure 9.15 Choose basic characteristics for the first button

2. Release the mouse to open the Button Properties dialog box, also seen in Figure 9.15, and display the General tab. Acrobat 6 and 7 both name the object *Button1* by default. Click the Form Field drop-down arrow and choose Visible but does not print from the list. This means the button is visible onscreen, but printing a page of the manual doesn't include the image of the button.

3. Click the Appearance tab to choose visual options for the button (Figure 9.16). Choose a color for the fill and border of the button, as well as its font, font size and font color. In the example, the button is gray, without a border, and uses white text.

4. Click the Options tab, used to add visual elements to the button, such as the text characters or a graphic element (Figure 9.17). Choose a combination of a Label (text) and/or an Icon (graphic image PDF) to use for the button. You can also specify the button's behavior. The default Up state is used in this example, which means the program responds to the user clicking and releasing the mouse, known as the *trigger*, over the button.

5. Click the Actions tab. The default trigger shown in the Select Trigger field is MouseUp, meaning that the action is activated when the user clicks and releases the mouse over the button (Figure 9.18). Choose the action from the Select Action drop-down list, which is the same as that offered for links. In the example, the Go to a page view action is selected.

6. Click Add; you see the same dialog box as that shown in Figure 9.12. Display the desired magnification, page, and location of the Table of Contents document. Click Set Link to close the dialog box and return to the Button Properties dialog box.

7. Click Close to dismiss the Button Properties dialog box and complete the button.

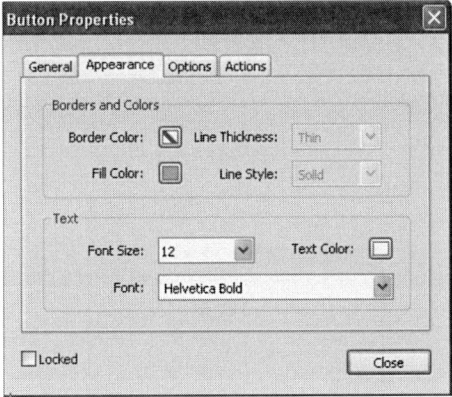

Figure 9.16 Choose visual characteristics for the button

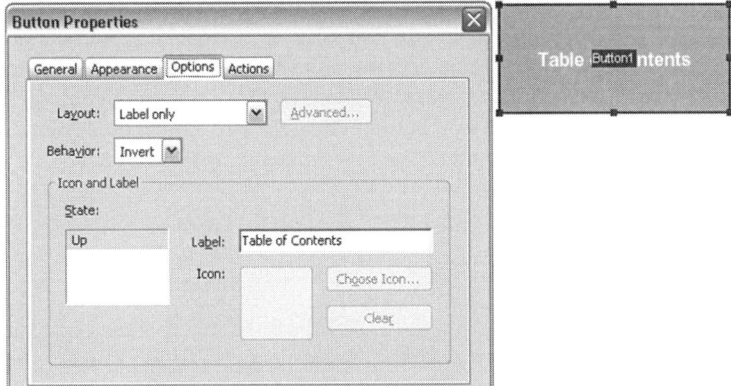

Figure 9.17 Select button options, such as icons, labels, or behavior

Note: Choosing any of the other Behaviors, except for a Push button, uses a single Up state as a trigger. The Push button uses a three-state button which can be configured for each state, using separate triggers when the user moves their mouse over, away from, or clicks and releases the mouse over the button.

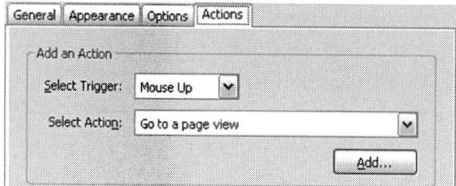

Figure 9.18 Choose the appropriate trigger and action for the button

Duplicating the Button

You can see from the example that building a single button is quite similar to adding a link. The difference lies in the ability to reuse the button easily, as you see in this section.

To create duplicates of your button throughout all or part of a document, follow these steps:

1. Right-click the button with the Button tool and choose Duplicate from the shortcut menu to open the Duplicate Field dialog box.

2. Click All to duplicate the button throughout a document, or click the From radio button and specify a page range, as in Figure 9.19.

3. Click OK to close the dialog box and add the duplicate buttons to the pages specified.

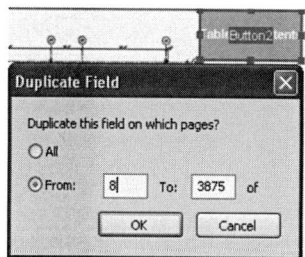

Figure 9.19 Select the page range for duplication

Acrobat handles a wide range of page sizes. As a result, it may be necessary to move the buttons on some pages and possibly make the buttons larger or smaller. To move a button, click on the button with the Button tool and drag in any direction; to resize a button, click a resize handle on any side of the button with the Button tool and drag to reconfigure the size.

Summary

Acrobat has a great set of easy-to-use navigation tools. Make sure to incorporate a navigation plan into your project design. Using appropriate navigation makes it easier for users to work with your project, and you can plan the most useful method for them to make their way through your document.

In this chapter you learned about using links, bookmarks, and buttons for navigation in a document. You saw how bookmarks can be added and modified, and how you can create bookmarks from the document's tags. You learned how to add a link and how to set a specific view that the user sees when the link is used. Finally, you saw how to create and modify a button to use for navigation, and how to create duplicates to easily control one type of navigation in a large file.

Exercises

1. Using a sample file, consider different ways to add user navigation through the file. Which is the most efficient way of providing navigation? Why did you choose one method over another method?

2. Experiment with using bookmarks in both a tagged and an untagged document. Use the tagged document's content to create bookmarks as described in the chapter.

3. Add and configure some navigation links in a sample document, experiment with using different link appearances and actions.

4. Use a sample button in a project file. Experiment with different button appearances and options.

Project

Using the project files, you will add a set of bookmarks for navigation, as well as a number of links and a button on each page that displays the Sheet Index page when clicked. Use the **DuPont Plan Set.PDF** file from the **ch09_project** folder, which is like a paper plan set. The PDF version does not contain any navigation features that make PDF files more user friendly. Bookmarks added in the PDF creation process are not useful and need to be customized.

Task 1: Setting the initial view

1. Open the **DuPont Plan Set.pdf** file.
2. Choose File > Document Properties > Initial View.
3. Choose Bookmarks Panel and Page from the Initial View drop-down menu.
4. Choose Single Page from the Page Layout drop-down menu.
5. Choose Fit Page from the Magnification drop-down menu; click OK to close the dialog box.
6. Save the file. Close it and reopen it – you should see the Bookmarks pane and one complete page display when the file opens.

Task 2: Revising the bookmarks

1. Click through the existing set of bookmarks, checking their order in the Bookmarks pane, their page order, and the magnification.
2. Rename each bookmark that corresponds with a page in the file. You can click the existing name's text to select it and then type a new name.
3. Delete extraneous bookmarks from the Bookmarks pane.
4. Drag the bookmarks into the correct order, corresponding with the page display in the Document pane.

Task 3: Adding a link from each sheet

1. Zoom into the Sheet Index on Sheet 1: a magnification of approximately 75% shows the index's text clearly.
2. Click the Link tool on the Advanced Editing toolbar to select it.
3. Drag a marquee around the C-2 Existing Conditions Plan label. Release the mouse to open the Link Properties dialog box.
4. Choose Invisible Rectangle from the Link Type drop-down menu on the dialog box.
5. Click the Go to a page view radio button in the Actions section of the dialog box.
6. Click Next to close the dialog box and open the Create Go to View instruction dialog box.

7. Click the bookmark in the Bookmarks pane that corresponds to the label enclosed in the link box to display the page in the Document pane.
8. Click Set Link to close the Create Go to View dialog box and set the link.
9. Repeat with the other five sheets in the document, and save the file.

Task 4: Adding a button for navigation

1. Display page 1 in the Document pane.
2. Choose the Button tool from the Forms subtoolbar on the Advanced Editing toolbar and drag a small marquee at the top right of page 1. Release the mouse to complete the marquee and open the Button Properties dialog box.
3. On the General tab, choose Visible but doesn't print from the Form Field drop-down menu.
4. On the Appearance tab, choose Border Color, none; Fill Color, gray; Text Color, white; Font Size, Auto.
5. On the Options tab, click the Label field and type SHEET INDEX.
6. On the Actions tab, select Go to a page view from the Select Action drop-down menu, and click Add to open the Create Go to View instruction dialog box.
7. Display a zoomed view of the Sheet Index on page 1 and then click Set Link to close the Create Go to View dialog box and set the action.
8. Click Close to dismiss the Button Properties dialog box and complete the button.
9. Click the Hand tool on the Basic toolbar, and click the button to test it.

Task 5: Duplicating the navigation button

1. Select the Button tool on the Advanced Editing toolbar.
2. Right-click the existing SHEET INDEX button to open the shortcut menu.
3. Choose Duplicate to open the Duplicate Field dialog box, and choose All to place duplicate buttons on each page of the document.
4. Check the location of the buttons on each page, moving them as required to prevent covering text.
5. Save the file.

10

Security

Maintaining the integrity and security of engineering drawings is a major concern to all parties involved in a project. The fear of unauthorized changes to a drawing made after issue has made many an engineer lose sleep.

Architecture and engineering law requires that drawings be sealed and signed by the engineer in charge of producing the drawings. In the digital world, an image of a seal and a signature is simply an image; anyone with a simple scanner can easily scan and reuse such an image for a set of plans.

Regulations Covering Electronic Documents

Many authorities and regulating bodies have issued requirements for electronic drawings. One of the best examples we have found of requirements for electronic drawings are the Regulations of the Board For Architects, Professional Engineers, Land Surveyors, Certified Interior Designers & Landscape Architects of Virginia. These regulations were developed in conjunction with the Navy Facilities Engineering Command (NAFAC) in Norfolk, Virginia: NAFAC has an excellent implementation of PDF in AEC.

18 VAC 10-20-760. Use of seal

The pertinent section of the Virginia Regulations follows:[6]

B.1.a. An electronic seal, signature and date is permitted to be used in lieu of an original seal, signature and date when the following criteria, and all other requirements of this section, are met:

1.	It is a unique identification of the professional;

2.	It is verifiable;

3.	It is under the professional's direct and sole control;

---

[6] Virginia Department of Professional and Occupational Regulation (2002) STATUTES, Title 54.1, Chapter 4, Excerpts from Title 13.1 pp. 48 – 49

4.   It is linked to the document file in such a manner that changes are readily determined and visually displayed if any data contained in the document file was changed subsequent to the electronic seal, signature and date having been affixed to the document: and

5.   Changes to the document after affixing the electronic seal, signature and date shall cause the electronic seal, signature and date to be removed or altered in such a way as to invalidate the electronic seal, signature and date.

B.1.b. In addition, once the electronic seal, signature and date are applied to the document, the document shall be in a view-only format, if the document is to be electronically transmitted.

As you will see in this chapter, the security and digital signature capabilities in Acrobat are able to meet all of the regulation's requirements.

In this Chapter

Acrobat 5 and 6 include both Password and Certificate Security; Acrobat 7 has added the ability to use the Adobe Live Cycle Policy Server. In this chapter you will learn how to work with a range of security methods:

* Password Security is the most familiar method of securing a document. In Acrobat, you can create two levels of passwords for controlling access to both opening and modifying a document.
* Certificate security is based on digital signatures created in Acrobat. In addition to creating certificates, you can also exchange them, set specific rights, and examine their contents.
* Certificate security can be applied in two ways, either as a signature, or as a certifying signature. Both signature options use the same process and encryption method; the difference lies in the order of signing, as only the first signature on a document can be defined as a certifying signature.
* Digital signatures can be customized, using combinations of the signature elements provided, optionally accompanied by a custom graphic image.
* Rather than recreating a set of parameters to apply with a signature or password each time it is used, you can create a policy that is stored in Acrobat and available for use as needed.
* Document attachments can be encrypted separately from the rest of a document using an eEnvelope.
* File versions based on existing content at the time of signing can be compared.

Using Password Security

Most forms of security used in Acrobat are based on permissions. That is, a user is granted the right to open a document and make allowable modifications. Unlike other security options available in Acrobat, using password security can protect the contents of a file against changes anonymously. You can also control access to the document using a Document open password.

Applying Password Security

There are several approaches to applying security in Acrobat; the simplest method to add a simple password is through the Document Properties dialog box.

Follow these steps to apply password security to a document:

1. Choose File > Document Properties > Security to open the Document Properties dialog box.

2. Click the Security Method drop-down arrow and choose Password Security from the list (Figure 10.1). The Password Security-Settings dialog box opens.

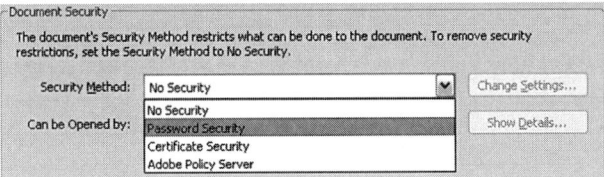

Figure 10.1 Select the security method in the dialog box

3. Choose a Compatibility option from the Compatibility drop-down list; a set of radio buttons become active (Figure 10.2).

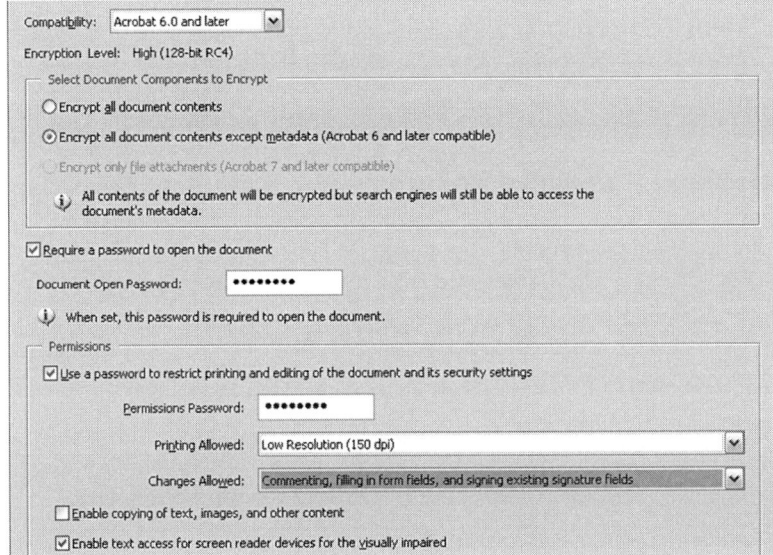

Figure 10.2 The compatibility option selected determines the available options

4. Select encryption options based on the selected Compatibility version. Refer to Table 10.1 for available options.

5. Click Use a password to restrict printing and editing of the document and its security settings; the Permissions Password field becomes active; type a password in the Permissions Password field.

6. Click the Printing Allowed and Changes Allowed drop-down arrows and select options from the lists.

7. Click OK to close the dialog box. A message displays describing the effect of third-party security products; click OK to close the message.

8. A Confirm Permissions Password dialog box opens; type the password in the Permissions Password field to verify the password and click OK. The confirmation dialog box closes, returning you to the Document Properties dialog box.

9. Click OK to close the Document Properties dialog box; click OK to close the message dialog box that states you have to save the document in order to save the security settings.

10. Click OK to dismiss the message and return to the program.

Often, saving a document using an alternate name makes it easier to work with the document in the future: it is simpler to make changes to a document and then save with password protection than it is to remove existing protection before making changes and reapplying the security again.

After saving, close and reopen the file to test the protection. Based on the selected security settings, such as not allowing printing, you will see that many of the program commands are disabled. A protected document displays a security icon at the lower left of the program window. Move your cursor over the icon to display a tooltip explaining the security applied (Figure 10.3).

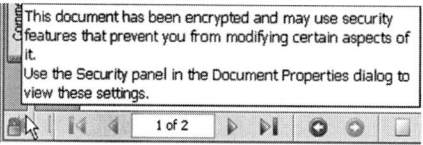

Figure 10.3 A protected file displays a security icon

Removing or Editing Password Security

Password protection can be removed simply from a protected document.

With the document open in Acrobat, follow these steps to remove the password security:

1. Choose File > Document Properties to open the Document Properties dialog box, and then select the Security tab.

2. Click the Security Method drop-down arrow and choose No Security from the list; a confirmation dialog box opens.

3. Type the password in the confirmation dialog box, and click OK to confirm the deletion and close the dialog box; click OK again to dismiss the Document Properties dialog box.

4. Save the file to preserve the security changes.

To edit existing protection, in the Document Properties dialog, click Edit Settings to open the same options used to apply the initial document passwords, make modifications as required, and resave the file.

Digital Signatures

Just as your signature on a document attests to your understanding and acceptance of the content you have signed, a digital signature can be applied to a document in the same way. A digital certificate is akin to a signature card at the bank: you sign the card and the bank uses it to verify the signature on a check.

Your signature in the digital world is a unique mathematical equation. To be valid in commerce, a third party must keep a copy of your Public Key to verify your signature. There are several companies, such as Entrust, RSA and VeriSign, that issue and manage digital signatures.

There are additional ways to apply security, of course. For instance, the US Navy is using a digital ID that also has a biometric component, and analyses a signature written on a signature pad.

Certificate Security is more powerful than password security, and less likely to be easily broken. Certificate Security is not password dependent, although you use a password associated with the digital signature when applying it to a document.

A document can be certified or digitally signed by application of a digital signature which remains valid unless the document is modified in ways other than those specified at the time of signing.

Table 10.1 Encryption options vary according to the compatibility version chosen

Version/ Encryption Level	Document Components Affected	Printing Permissions	Change Permissions	Other Permissions
Acrobat 3.0, 40-bit RC4	All contents	None, high resolution	Commenting, filling in form fields, and signing existing signature fields; page layout, filling in form fields, and signing existing signature fields; any except extracting pages	Enable copying of text, images, and other content and access for the visually impaired
Acrobat 5.0, 128-bit RC4	All contents	None, low resolution, high resolution	Inserting, deleting, and rotating pages; filling in form fields and signing existing signature fields; commenting, filling in form fields, and signing existing signature fields; any except extracting pages	Enable copying of text, images, and other content; enable text access for screen reader devices for the visually impaired
Acrobat 6.0, 128-bit RC4	All contents; all document contents except metadata	None, low resolution, high resolution	Inserting, deleting, and rotating pages; filling in form fields and signing existing signature fields; commenting, filling in form fields, and signing existing signature fields; any except extracting pages	Enable copying of text, images, and other content; enable text access for screen reader devices for the visually impaired
Acrobat 7.0, 128-bit AES	All contents; all document contents except metadata; only file attachments	None, low resolution, high resolution	Inserting, deleting, and rotating pages; filling in form fields and signing existing signature fields; commenting, filling in form fields, and signing existing signature fields; any except extracting pages	Enable copying of text, images, and other content; enable text access for screen reader devices for the visually impaired

Creating a Digital Signature

Acrobat contains two complementary processes called *certifying* and *signing*. The difference between the two lies in their application. The document's originator can either certify or sign a document; subsequent digital IDs can be applied as signatures only.

The process of creating the digital signature used for both signing and certifying a document is the same. The visual signature applied to a file, called an *appearance*, can use default characteristics or a custom signature appearance, such as a seal or a graphic image.

A digital ID is composed of a public and a private key, both of which are generated by Acrobat when you create the digital ID. The private key is maintained on your system and used for signing and certifying documents; the public key is used for exchanging certificates with others.

Specifying Digital Signature Characteristics

The terms *digital signature*, *Digital ID*, and *digital profile* all refer to the same digital identification used to certify or sign a document, and can be used interchangeably.

To create a new Digital ID, follow these steps:

1. Choose Advanced > Security Settings to open the Security Settings dialog box.

2. Click (+) to the left of the Digital IDs in the left frame of the dialog box to display any existing ID files in the upper-right frame of the dialog box (Figure 10.4).

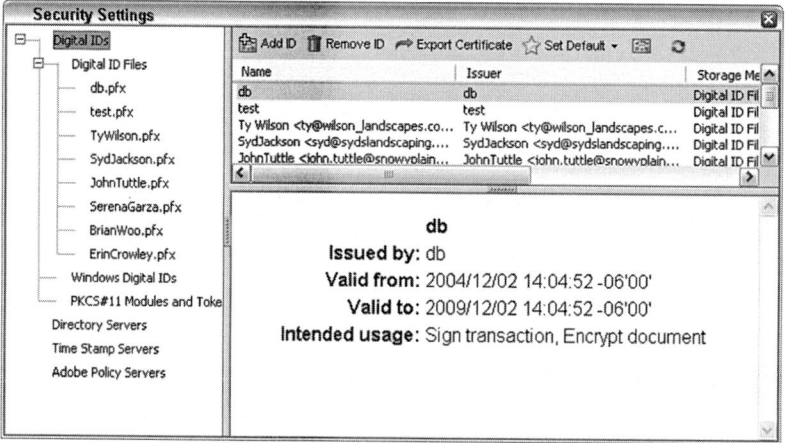

Figure 10.4 Manage digital IDs in the Security Settings dialog box

3. Click Add ID on the dialog box's toolbar to open the Add Digital ID dialog box, which offers three options: find an existing ID, create a new ID, or retrieve one from a third-party source.

4. Click Create a Self-Signed Digital ID; click Next at the bottom of the dialog box.

5. The next pane displays information about exchanging certificates with others to use the selected type of security; click Next at the bottom of the dialog box.

6. Choose a storage type for the Digital ID in the next pane. The two options available are a PKCS#12 Digital ID file, which is the default selection, or add the Digital ID to your Windows Certificate Store. Click an option – the example is a PKCS#12 Digital ID file – and then click Next.

7. Add information to include in the certificate in the next pane of the dialog box, including your name and other identifying information.

8. Choose a Key Algorithm from the drop-down list, and choose how the Digital ID will be used – either for digital signatures, data encryption, or both (Figure 10.5). Click Next.

9. In the final pane of the dialog box, specify the certificate's location on your hard drive or server. The default storage location is Acrobat's Security subfolder, part of the program installation. Click Browse to choose a different storage location.

10. Type and confirm a password used to apply the certificate when used. The password must be at least six characters in length.

11. Click Finish to close the dialog box and save the new Digital ID. Acrobat stores the Digital ID in the selected folder. You will see the new certificate added to the list in the Security Settings dialog box.

12. Click OK to close the Security Settings dialog box.

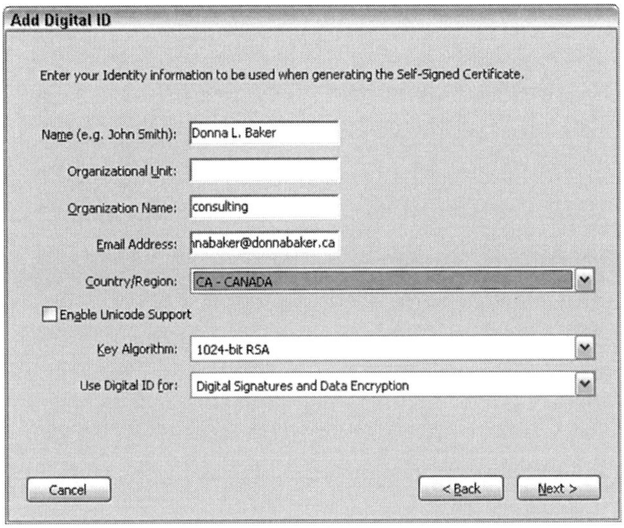

Figure 10.5 Specify a key algorithm and identifying information

Security Options

You can choose compatibility versions from Acrobat 3.0 to 7.0, which offer different levels of encryption. Acrobat versions 3.0 – 6.0 use RC4 security algorithms at different bit values. RC4 is a stream cipher produced by RSA Security. It uses an algorithm based on the use of a random permutation.

Acrobat version 7.0 offers the AES (Advanced Encryption Standard) method of security, a 1024-bit encryption algorithm that can provide different levels of security for a document or just its attachments. It is the encryption method presently used for securing sensitive but unclassified material by US Government agencies.

Acrobat 4.0-compatible documents can be viewed by the most users, Acrobat 7.0-compatible documents by the least number of viewers.

Customizing the Signature Appearance

Rather than using the basic appearance, a signature can be customized and then selected when the signature is applied to the document.

Follow these steps to create a custom signature appearance using a graphic image:

1. Choose Edit > Preferences to open the Preferences dialog box; then choose Security from the column on the left. The Appearance window at the top of the dialog box lists existing signature appearances. Click New.

2. The Configure Signature dialog box opens. Enter a name for the appearance.

3. Choose a Configure Graphic option, either your default name, no graphic, or an image from a file (Figure 10.6). The example uses an image.

4. To use an image, click the Imported Graphic radio button, then click File to open the Select Picture dialog box.

5. Click Browse to open a dialog box to locate and select the image you want to use. Click Select to close the Browse dialog box and load the image in the Select Picture dialog box (Figure 10.7).

6. Click OK to close the dialog box; the selected image is shown at the top of the Configure Signature Appearance dialog box.

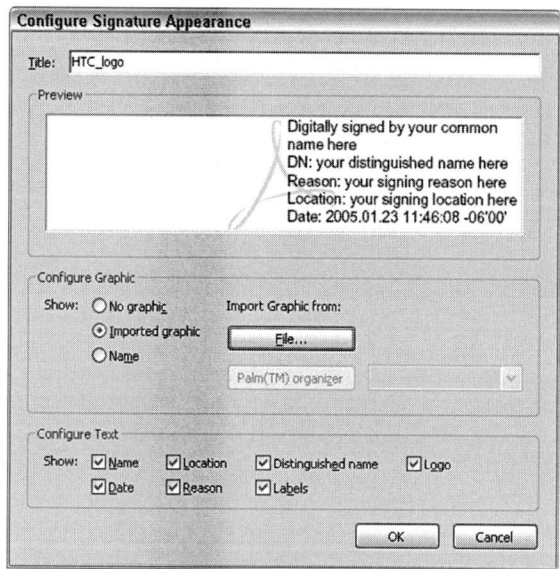

Figure 10.6 Name and choose options for creating a custom signature appearance

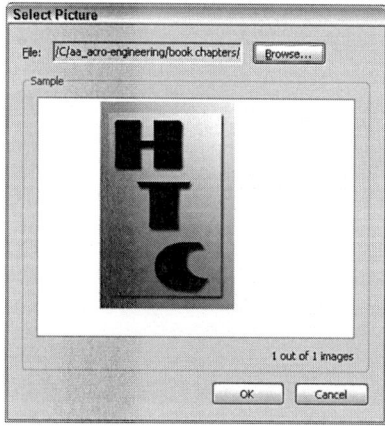

Figure 10.7 The selected image displays in the preview dialog

7. Specify text options in the Configure Text section of the dialog box. All options are selected by default.

8. Click OK to close the Configure Signature Appearance dialog box and add the new signature appearance to the Appearance list on the Preferences dialog box.

9. Click OK to close the Preferences dialog box.

When you apply a signature to a document, you can select any custom signature appearance stored on your system.

Certifying a Document

If you are the originator of a document, the first signature applied to secure the document can be a certifying signature. Before certifying a document, make sure all modifications and changes are made to the file, as making changes after certification can corrupt the signature.

Follow these steps to certify a document using a custom appearance:

1. Choose File > Save as Certified Document to open an information dialog box, explaining that certification vouches for the contents of the document and that you need a digital ID to proceed. Click OK to close the information dialog box.

Note: If you intend to use a third–party certification system, click Get Digital ID From Adobe Partner, and proceed through the prompts.

2. The Save as Certified Document – Choose Allowable Actions dialog box opens. Choose an option from the Allowable Actions drop-down list. The options include commenting and form fill-in actions, no changes, or only allow form fill-in actions (Figure 10.8).

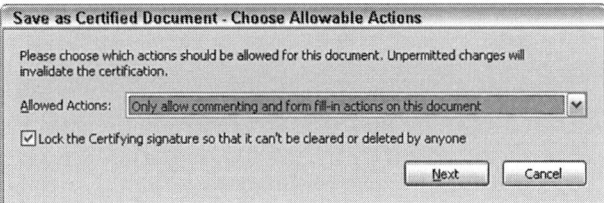

Figure 10.8 Define the user actions allowed in the certified document

3. To prevent anyone from trying to remove the signature, click the Lock the Certifying signature so that it can't be cleared or deleted by anyone checkbox.

4. Click Next; the Save as Certified Document – Warnings dialog box opens. A list of items that can jeopardize the integrity of the certification are listed. In the example shown in Figure 10.9, the document contains active URLs as well as allowing commenting on the document. A default warning is selected on the dialog, which explains that the content is used to enhance the document's interactivity. When the recipients open the document, they see both the warning and the selected comment.

5. Click Next to display the Save as Certified Document – Select Visibility dialog box. Choose either to show or hide the certifying signature on the document.

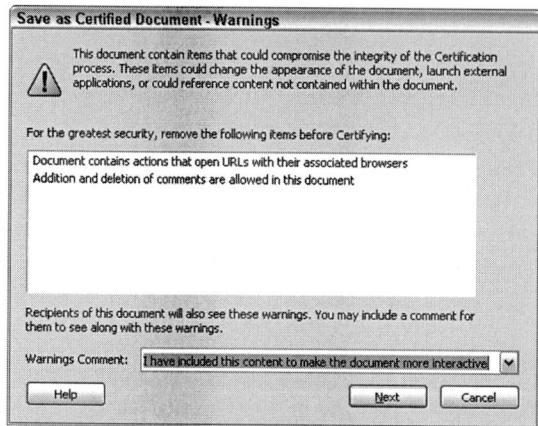

Figure 10.9 Read the warnings resulting from allowable user actions

6. Click Next to display an information dialog box describing how to draw the signature field; click OK to close the dialog box.

7. The Signature tool is activated; draw a marquee on the page to hold the certifying signature and then release the mouse to complete the marquee.

8. The Apply Digital Signature – Digital ID Selection dialog box opens. Select the signature to use for the certification and cllick OK to close the dialog box.

9. The Save as Certified Document – Sign dialog box opens (Figure 10.10).

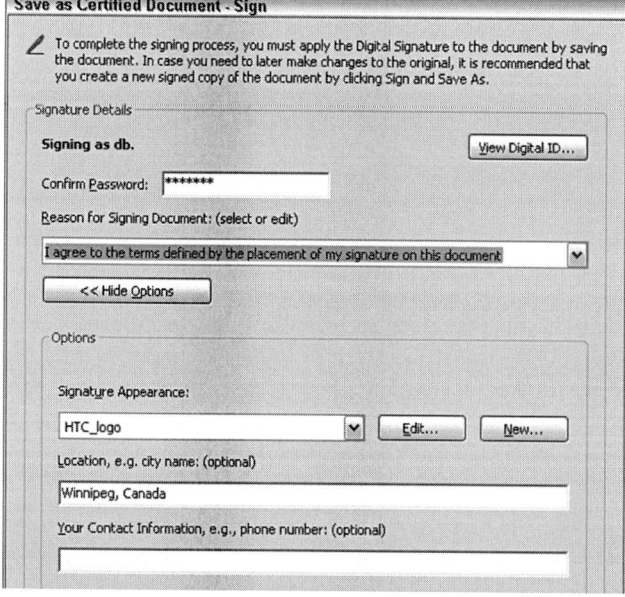

Figure 10.10 Select options to customize the appearance of the certifying signature

In this dialog box:

- Type the Digital ID's password in the Confirm Password field.
- Click the Reason for Signing Document drop-down arrow and choose an option from the list. The options range from attesting to the integrity of the document to approving the contents.
- Click Show Options to expand the dialog box. In this area, you can choose a custom appearance from the Signature Appearance drop-down list, as well as add location and contact information. The Show Options/Hide Options buttons toggle the visibility of the lower portion of the dialog box.

10. Select a save option at the bottom of the dialog box – either Sign or Sign and Save As. Name the file and click Save to close the dialog box and save the file.

On the document, you see the certifying signature and options chosen in the dialogs such as the reason for certifying (Figure 10.11).

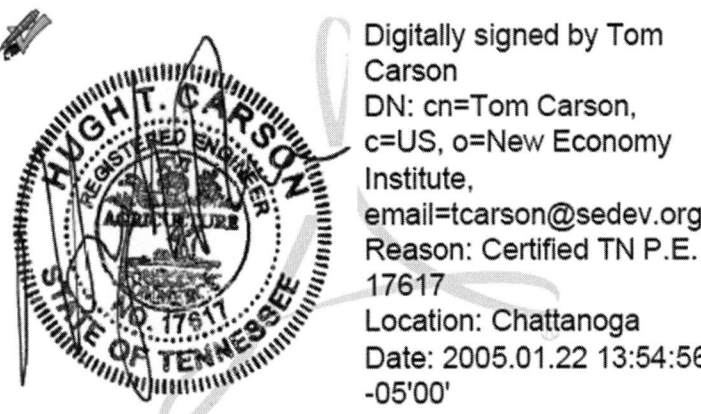

Digitally signed by Tom Carson
DN: cn=Tom Carson, c=US, o=New Economy Institute, email=tcarson@sedev.org
Reason: Certified TN P.E. 17617
Location: Chattanoga
Date: 2005.01.22 13:54:56 -05'00'

Figure 10.11 The certifying signature displays on the page

Note: The border of engineering plans usually displays a square just big enough for the seal. The area for a digital signature may need to be larger to accommodate the signature.

Finding Certification Information

When a certified document is opened in Acrobat or Adobe Reader, a Certified Document icon 🔒 is shown at the left of the status bar. You can find information about the certificate used to certify a document, including details of the certifier, and the encryption used from the document.

Use one of these methods to find certification information:

- Hold the cursor over the icon to display first a simple tooltip and then a more descriptive tooltip showing signature details (**Figure 10.12**).
- Click the Certified Document icon to open a dialog box describing the status of the document, security features, and access buttons to click for more legal and signature information.
- Choose View > Navigation Tabs > Signatures to open the Signatures pane and then click (+) to the left of the signature listing to read details about the certifying signature.

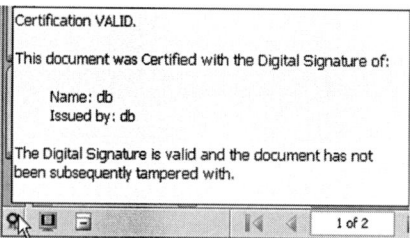

Figure 10.12 Read signature information in a tooltip

Adding Signature Fields

A signature field can be added when someone signs a document, or a signature can be placed into an existing signature field on a document. You may want to add blank signature fields to a document if you are the originator of the document, for example, and need approval signatures from a number of others. A certified document cannot be signed by anyone else unless there are blank signature fields added to the document prior to certification.

Follow these steps to add a signature field to a document:

1. Click the Sign task button's drop-down arrow and choose Create a Blank Signature Field from the menu.

2. An Adobe Acrobat dialog box opens saying the Signature Field tool is selected and that you can draw a marquee to place the field; click OK to close the information dialog box.

3. Draw a marquee on the page for the signature field; release the mouse to open the Digital Signature Properties dialog box, which displays the Appearance tab by default.

4. Select visual options for the signature field, such as border and fill colors and styles, on the Appearance tab.

5. Click the General tab; the field is named Signature1 by default. Leave the default text or type custom text.

6. Click the Tooltip field and type text instructions if desired.

7. Click Close to dismiss the dialog box and complete the field.

Tip: If you are planning to using Acrobat's automatic field duplication process, delete the "1" suffix from the default name to make use of the auto naming feature.

The blank signature field is shown on the document page in the area specified by the marquee (Figure 10.13). It is also listed in the Signatures pane. If you add a signature to a document rather than signing a blank signature field, the signature can be further protected.

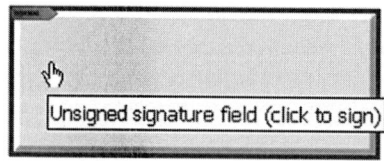

Figure 10.13 Add blank signatures to a document before routing for approval

Making a Signature Field Read-only

In many circumstances, it is important that a signature field added to a document is converted to a read-only field after signing.

You can define the action for the field by following these steps:

1. Follow the previous set of steps to draw and configure the field.

2. Click the Signed tab in the Digital Signatures Properties dialog box.

3. Click the Mark as read-only radio button, and choose Just these fields from the drop-down list.

4. Click Pick to open a list of the fields in the Field Selection dialog box.

5. Select the fields you want to define as read-only and click OK to close the dialog box. The fields selected are listed on the Digital Signature Properties dialog box (Figure 10.14).

6. Click OK to close the Digital Signature Properties dialog.

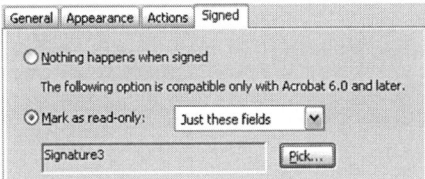

Figure 10.14 Define a signature field as read-only to protect it after signing

Signing a Document

Blank signature fields are useful for an approval cycle, for instance, where a document is reviewed by several people, each signing to specify their acceptance of the contents. If you need to protect a signature further, you can sign a document and preserve the signature in a read-only field.

Follow these steps to sign and preserve a signature in a document that contains no other signatures:

1. Click the Sign task button's drop-down arrow and choose Sign This Document.

2. The Document is Not Certified dialog box opens, giving you the option to continue with signing, or certify the document instead. Click Continue Signing to close the dialog.

3. The Sign Document dialog box opens. Choose either to display the signature or create an invisible signature field, and click Next.

4. The dialog box closes, and the Signature tool is active. Draw a marquee on the document to define the signature's location.

5. Release the mouse; the Apply Digital Signature – Digital ID Selection dialog box opens. Select the digital ID to use for the signature (Figure 10.15).

6. Select a Digital ID persistence option on the dialog box. You can choose from three levels of persistence: always use the same ID, use the same ID for the current program session, or ask each time a digital ID is applied.

7. Click OK to select the signature and display the Apply Signature to Document dialog box (identical in content to the Save as Certified Document - Sign dialog box shown in Figure 10.10).

8. Select signature options, and then click Sign and Save, or Sign and Save As to apply the signature and save the document.

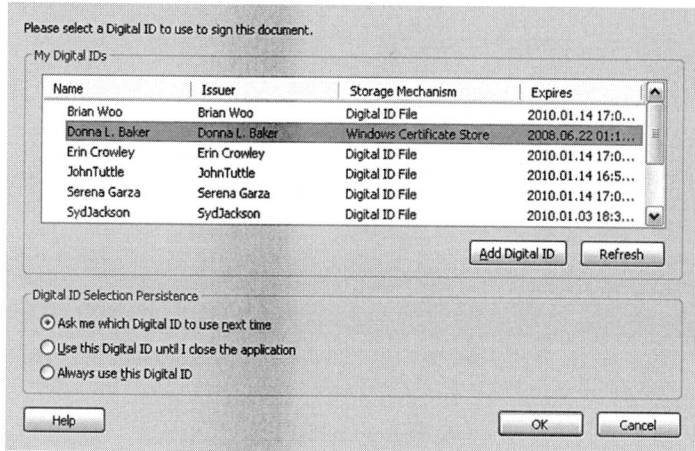

Figure 10.15 Choose a digital ID and its persistence in the dialog box.

Sharing and Extracting Digital IDs

A signed or certified document uses the signator's Private Key from their digital certificate. In order to exchange encrypted content with others, you need to exchange certificates, called *Trusted Identities*. Certificates can be exchanged by email, or extracted from existing signatures. The most secure method is to exchange certificates.

Tip: Contacts and Certificates options are similar; a Contact lists the certificate by the certificate-holder's name.

Exchanging Certificates

Certificates are exchanged in FDF (file data format) files. You can attach a certificate to an email as you would any other type of file, or use Acrobat's data file exchange function, which allows you both to send a certificate to a contact and receive their certificate in return.

Follow these steps to exchange certificates in Acrobat:

1. Choose Advanced > Trusted Identities to open the Manage Trusted Identities dialog box and select either Contacts or Certificates from the Display drop-down list (Figure 10.16 shows a Certificate listing). Click Request Contact.

2. The Email a Request dialog box opens. Type your name and email address and optionally include a telephone number for the recipient to contact you to verify the request.

3. Click the Include my Certificates check box to automatically send your certificate to the recipient at the same time as you make your request, saving time in exchanging certificates. Click the Email request radio button to continue with the process; click Save request as a file to send it at a later time.

4. Click Next to display the Selecting Digital IDs To Export dialog box. Choose the certificate you want to send from the list, and click Next.

5. The Compose Email dialog box opens with a boilerplate subject and email content. Type the recipient's email address and click Email.

6. The content of the Compose Email dialog box is transferred to an email message.

7. Send the email, which includes the FDF file as an attachment (Figure 10.17). The boilerplate instructions describe what the certificate is and why the recipient is being sent the file. It also details that Acrobat 6 or 7, or Adobe Reader 6 or 7 is required in order to use the file attachment.

8. In Acrobat, click Close to dismiss the Manage Trusted Identities dialog box.

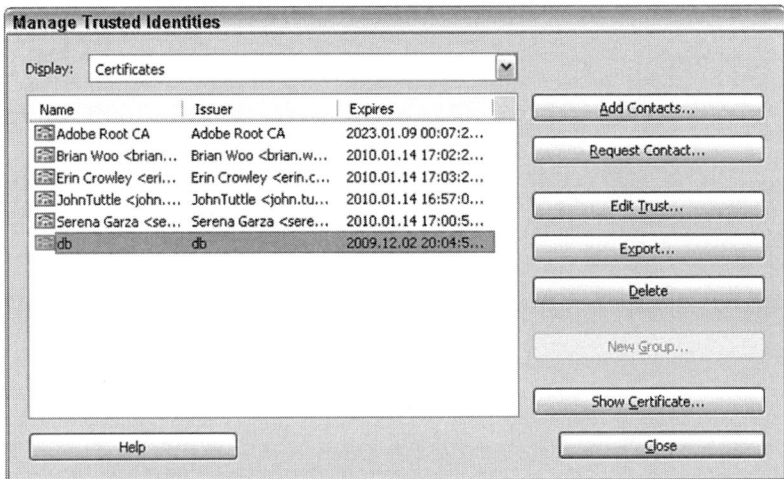

Figure 10.16 Export and import certificates through the Manage Trusted Identities dialog

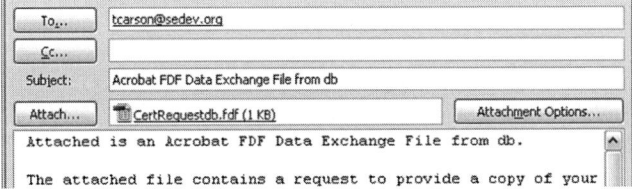

Figure 10.17 Acrobat automatically attaches the certificate file to the email

Extracting and Examining Certificates

Rather than emailing and requesting a certificate, you can extract it from an existing document. You can also review characteristics of an extracted certificate. The certificate information is exported from the information bedded within the signature. You can export it from the signature to store on your computer in several formats, including a PDF file or a certificate file.

Follow these steps to extract a certificate from a signed document and save it to a file as an Acrobat FDF Data Exchange file:

1. Click the signature in the document with the Hand tool 🖑 to open the Signature Validation Status dialog box, listing the validity, modifications, and whether the document is signed by the current user.

2. Click Signature Properties to open the Signature Properties dialog box; basic information is shown in the Summary tab, displayed by default (Figure 10.18).

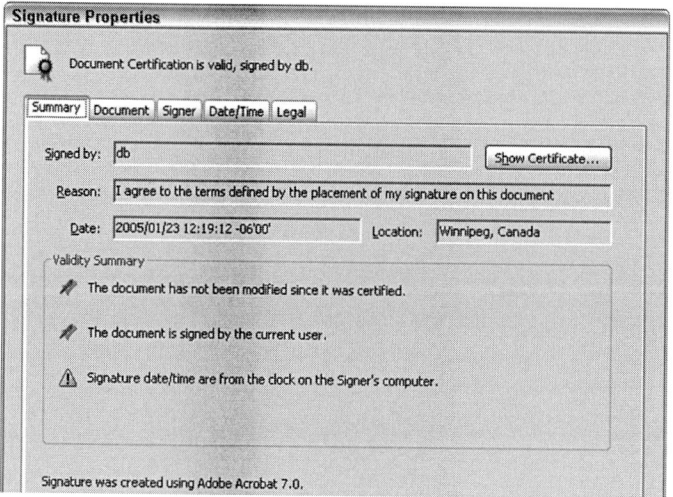

Figure 10.18 Read about certificates in the Signature Properties dialog box

3. Click Show Certificate to open the Certificate Viewer dialog box. Here you will find advanced information about the signature, such as legal status, encryption details, trust settings, and revocation dates. Certificate details are shown in Figure 10.19.

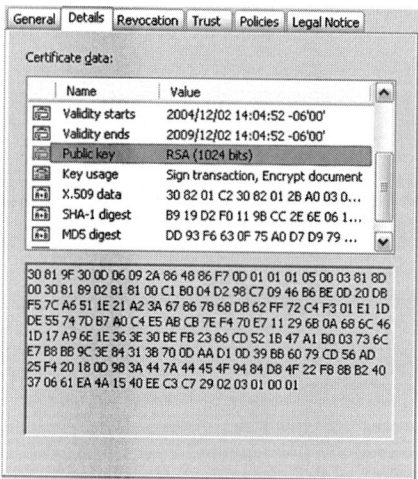

Figure 10.19 Read details about the certificate, such as its Public key information

4. Click the General tab on the Certificate Viewer dialog box and click Export; the Data Exchange File-Exporting Certificate dialog box opens. The dialog box is a three- or four-pane wizard.

5. Select a destination from the Choose Export Options pane of the wizard. The example uses the default destination and option; that is, Save the exported data to a file option, and Acrobat FDF Data Exchange (Figure 10.20).

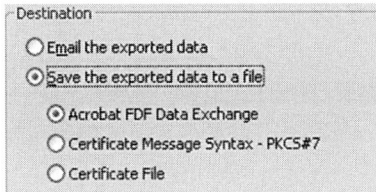

Figure 10.20 Choose a destination and type of file to export

6. Click Next to open the Sign Data Exchange File pane of the dialog, where you can digitally sign the data to exchange with the originator of the signature you are extracting.

7. Click Next to open the Choose a Path to Save File dialog box, and select a location for the file. Click Save to close the dialog box and display the selected path on the wizard dialog box.

8. Click Next to open a Review pane on the wizard showing information from the certificate, and click Finish to export the certificate and close the wizard dialog box.

9. Click OK to close the Certificate Viewer dialog box, and then click Close to dismiss the Signature Properties dialog box.

Importing a Certificate

Certificates are stored on your hard drive, and can be imported into Acrobat for future use in the Manage Trusted Identities dialog box. Follow these steps to import a certificate:

1. Choose Advanced > Trusted Identities to open the Manage Trusted Identities dialog box.

2. Click Add Contacts to open the Choose Contacts to Import dialog box.

3. Click Browse or Search to locate and select the certificate to open.

4. Click Open. The Locate Certificate File dialog box closes, and the file is listed in the Contacts area on the Choose Contacts to Import dialog box.

5. Click Import to close the Choose Contacts to Import dialog box and process the certificate file.

6. The Import Complete dialog box displays when the certificate is processed; click OK to close the dialog box, returning you to the Manage Trusted Identities dialog box.

7. The new certificate is now added to the list. Click Close.

Using Security Policies

Often, you find that you are repeating the same sequence of actions to apply consistent security settings to a series of documents or projects. Rather than manually applying the certificate or password to the document, create a security policy instead. You can build security policies for passwords, certificates, or Adobe Policy Server; the discussion is limited to passwords and certificates.

Creating a Password Security Policy

Password protection is commonly used to prevent changes in documents distributed anonymously, such as content intended for Web site use.

Follow these steps to create a new password security policy:

1. Click the Secure task button to open its menu and choose Manage Security Policies to open the Managing Security Policies dialog box.

2. Click New to open the New Security Policy dialog box, a four-pane wizard.

3. On the first pane of the dialog box, choose the basic encryption option you want to use for the policy. Leave the default Use passwords selection. Click Next.

4. In the General settings pane of the wizard, name the policy, and type a description. Click Save passwords with the policy to make it simpler to apply the policy at a later date (Figure 10.21).

5. Choose Document Restrictions in the next pane of the dialog. The contents of the dialog are the same as those shown in Figure 10.2.

6. Click Next; type the confirmation in the Adobe Security dialog, and click OK.

7. Click Next again to display the final pane of the dialog, which lists the Policy Details.

8. Click Finish to close the wizard and load the policy into the Managing Security Policies dialog.

9. Click Close to dismiss the dialog.

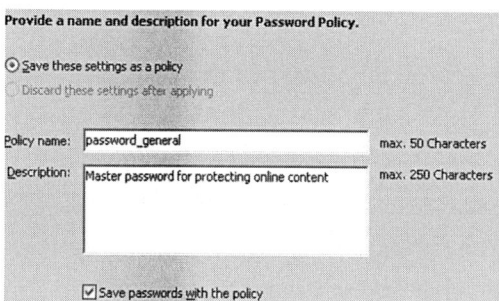

Figure 10.21 Use a meaningful name and description for a policy

Creating a Certificate Security Policy

In many workgroup situations you may find you are using the same policies in the same ways. Again, as with the password policy, creating a certificate security policy can save time in your workflow. In addition to choosing settings for the policy, you can also define the list of recipients.

Follow the same general instructions as those listed for a password-based policy. In the wizard, you will find different options from those listed for a password-based policy:

1. On the General settings pane of the wizard, in addition to naming and describing the policy, select the document components to be encrypted, as well as the Encryption Algorithm – you can choose either 128-bit RC4 or 128-bit AES encryption (Figure 10.22).

2. Click Next to open the Document Security-Digital ID Selection dialog, shown earlier in Figure 10.15. Select a digital ID and click OK to close the dialog, returning you to the wizard's third pane.

3. In the Select recipients pane of the wizard, list the recipients who receive the document secured using the policy. Use the options available from the wizard specify permissions.

4. Click Next to show the Summary pane of the wizard, listing the Policy Details.

5. Click Finish to close the wizard and return to the Managing Security Policies dialog.

6. Click Close to dismiss the dialog.

Working with Policies

If you habitually use the same policies, rather than opening the dialog boxes to select a policy each time, add it to the Favorites, which are then listed in the menu. Click the Secure task button to open the menu, and choose Manage Security Policies to open the dialog box. Select the policy to add to your favorites, and then click Favorites on the dialog's toolbar. A star displays to the left of the selected policy's name. Click Close to dismiss the dialog box.

The next time you need to apply the policy, click the Secure task button to open the menu, and you now find the favorite policy listed on the menu (Figure 10.23). Click the policy to select it from the menu, and then follow the dialog boxes to apply it to your document.

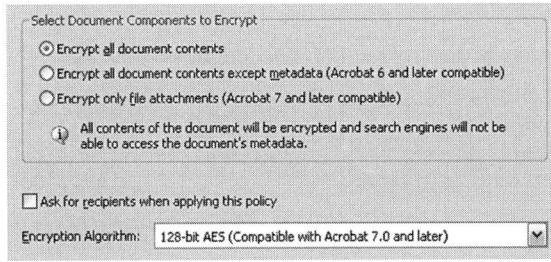

Figure 10.22 Choose an encryption algorithm and specify document components for encryption

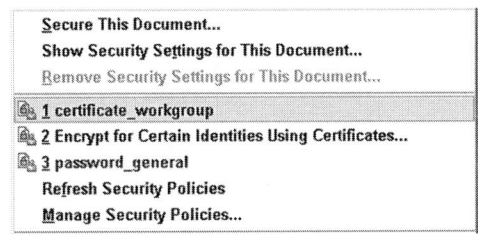

Figure 10.23 Add security policies to the menu for easy access

Secure PDF File Delivery

New in Acrobat 7 is the ability to encrypt just the attachments in a document in a secure wrapper called an eEnvelope. Acrobat includes a set of three default templates you can choose for securing the attachments (Figure 10.24). You can also use an image to create a custom template.

Encryption using an eEnvelope does not affect the attachments in any way: the process merely protects attachments from viewing by unauthorized users.

Open the file you want to secure, and then follow these steps:

1. Choose Document > Security > Secure PDF Delivery, or click the Secure task button and choose Secure PDF Delivery from the menu to open the Creating Secure eEnvelope wizard.

2. In the first pane of the wizard, click Add file to send to choose the documents to attach to the PDF, which also includes any attachments existing in the document. Click Next.

3. Choose a template from the second pane of the wizard, and click Next.

4. Select a delivery method for sending the file in the third pane of the wizard. The eEnvelope can be completed and emailed manually or automatically. The default option is to complete the template automatically, although none of the preconfigured templates can be sent automatically as they contain fields that need content added before sending. Select the manual option to add text in the template's fields after the wizard is complete. Click Next.

5. Click ShowAll Policies on the next pane of the wizard, and select the policy to use for the secure delivery (Figure 10.25). Click Next.

6. The final pane of the wizard shows a summary of your selections. Click Finished to close the wizard.

7. Depending on the security policy you selected, additional dialogs display for using a password or a certificate. Follow through the dialogs to secure the content.

8. The final step depends on the delivery option selected in the wizard:

   - If you chose to send the eEnvelope, enter your recipient's email address in the email dialog that opens and click Send.
   - If you chose to finish and email the eEnvelope manually, the wizard closes and the template is displayed in Acrobat. Fill in the fields as necessary, open your email program, and address and send the file.

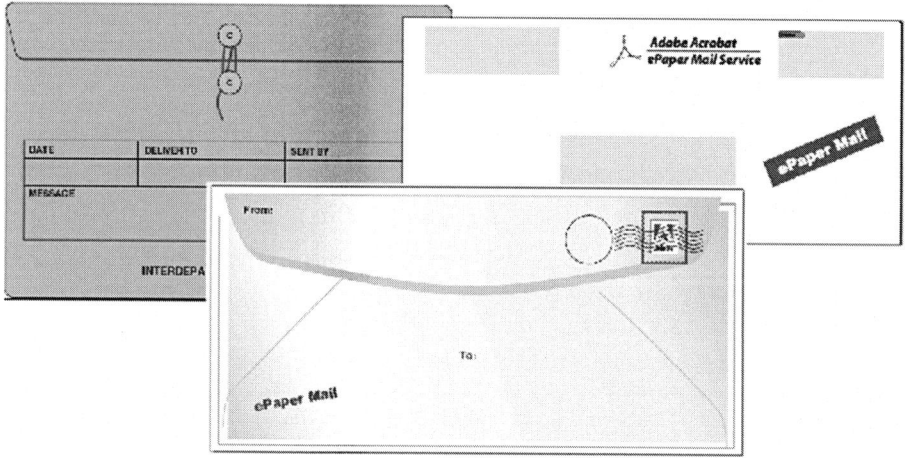

Figure 10.24 Acrobat 7 includes a set of default eEnvelope templates

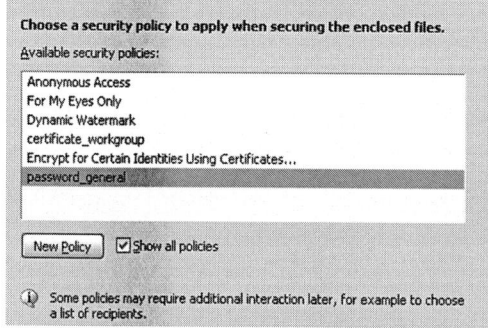

Figure 10.25 Select a policy to use for the eEnvelope

Adobe Live Cycle Policy Server

Adobe Live Cycle Policy Server is a security system that dynamically controls PDF document security. Adobe Policy Server is Web based and can be configured to run with various enterprise systems, including LDAP, and ADS server systems.

Storing, refreshing, and monitoring policy activity is controlled on the server; PDF documents are not stored on the server. Document users receive rights to use a document based on criteria applied through the server-based policies.

Follow these steps to log into Adobe Policy Server:

1. Choose Advanced > Security Settings to open the Security Settings dialog box.

2. Select Adobe Policy Servers from the policy type listing at the left of the dialog box.

3. Select a server on the right, and click Edit (Figure 10.26).

4. Type your user name and password, and then click Connect To This Server.

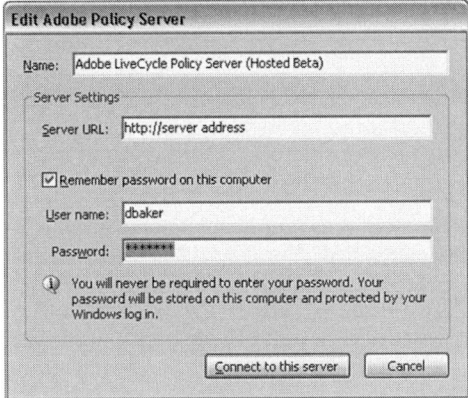

Figure 10.26 Select the security server from the dialog box.

You can view, monitor, and modify security policies (depending on the rights granted to you by the server administrator) by following these steps:

1. Click the Secure task button to display its menu, and choose Use APS Web Console.

2. In the Login screen that displays in your browser window, type your user name and password, and then click OK.

3. Once logged into the server, you can view and monitor organizational policies and policies you created through the browser interface.

Comparing Signed Document Versions

Regardless of whether you use certificate security or an Adobe Policy Server policy, you can compare signed versions of a document, or even compare two different documents. As each signature is added, Acrobat appends a copy of the document as it existed at the time of signing to the original document.

You can evaluate and compare the content of different versions of a document by following these steps:

1. Choose Document > Compare Documents to open the Compare Documents dialog box (Figure 10.27).

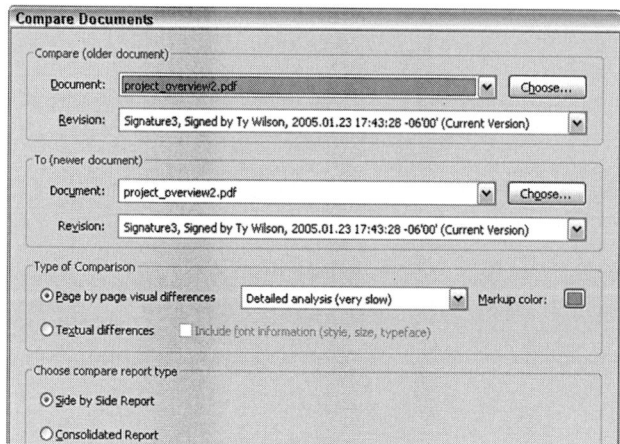

Figure 10.27 Choose criteria for comparing two document versions

2. Select the Document and Revision to use for comparison from the Compare (older document) drop-down lists; click Choose to select another file.

3. Select the Document and Revision to use for comparison from the To (newer document) drop-down lists; click Choose to select another file.

4. Select a type of comparison. You can use either visual or textual differences:

   - Page by page visual differences is selected by default. Choose Detailed, Coarse, or Normal from the drop-down list; Detailed analysis is selected by default. Click the color swatch to choose a Markup color from a color picker.
   - Textual differences compares the content in the two documents. You can also include differences in font information.

5. Select the type of report. You can use the default Side by Side Report, which compares the pages on alternating pages, or create a Consolidated Report.

6. Click OK to close the dialog and generate the report.

The report is opened in Acrobat as a temporary PDF file. Scroll through the pages to see the differences highlighted. The older version is shown at the left of the Document pane and the newer version at the right (Figure 10.28).

Using the Signature Pane Comparison Method

Rather than working through the Compare Documents dialog box, use the Signatures pane command. You do not have the same control over variables used in comparison, but it is a good method for a quick check.

   In the Signatures pane, select the signature you want to compare with the current version. Choose Compared Signed Version to Current Version from the Options menu in the Signatures pane. A temporary PDF file opens showing a page-by-page comparison of the document content highlighting the differences, as shown in Figure 10.28.

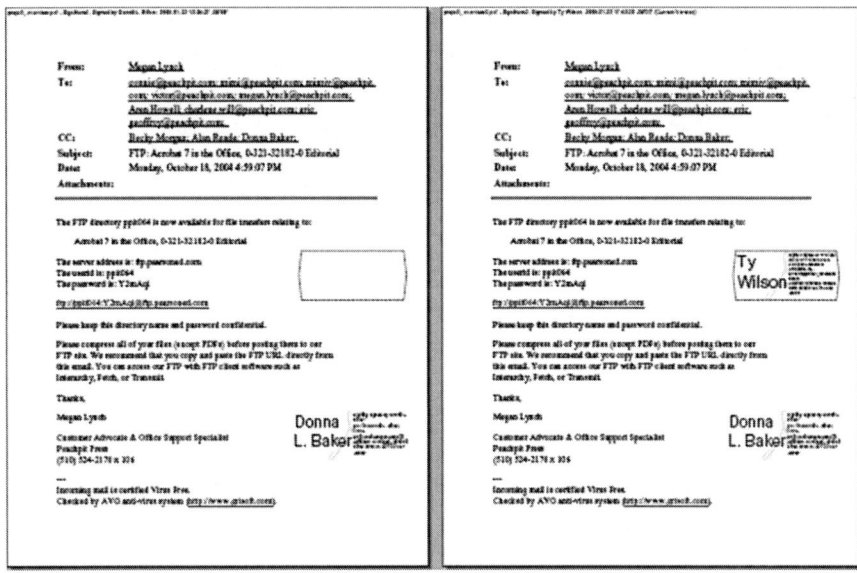

Figure 10.28 Comparing pages in two versions of a document side by side

Summary

In this chapter we examined Acrobat security, looking at the two most common types of security used in Acrobat – passwords and certificate security.

Passwords can be used for controlling access to a document, or to prevent changes to specified content. Certificate security uses digital signatures to control both the certification and the signing of a document for security purposes.

Acrobat includes several methods for applying signatures, verifying their content, as well as exchanging, extracting, and importing certificates. Digital IDs can be applied in a number of ways; their appearance on the document can be invisible, use standard content, or use a custom signature appearance.

You can manage and control Trusted Identities in a number of ways. To streamline your workflow, create policies that can be applied to a document or an eEnvelope. Regardless of the type of document being examined, and its signature status, you can compare different versions of the document.

Exercises

1. Using the method outlined in the chapter, construct and save a digital signature.

2. Use the signature constructed in Exercise 1 to digitally certify a document; use the signature to digitally sign a document.

3. Manipulate certificates – exchange certificates, extract a certificate from an existing signature, request a certificate by email.

4. Create a custom signature appearance.

5. Encrypt attachments to a document using an eEnvelope.

6. Using several versions of a signed document, compare the contents and generate a report.

Project

Using the project files, construct a digitally signed plan. The goal is to configure the signature fields to preserve the integrity of the documents, as well as to use a logical field-naming structure, following these steps:

1. Devise a signature field-naming structure. Here's an example: Sig.C01 (first page of Civil set), Sig.C02 (second Page of Civil Set), and so on; Sig.M01 (first page Mechanical), Sig.M02 (second page Mechanical), and so on.
2. Draw the signature fields. Place the fields at the location of the combined seal, signature and date on each page.
3. Specify that each signature field is set to read only once it is signed.
4. Sign the documents.
5. Apply password security to the Record set, disallowing any changes, but allowing printing and form fill-in and signing.

11

Using Acrobat Forms and Databases – Lyn Price

Everyone uses a form at some time or another, and many companies are using electronic forms to collect data in a more timely fashion. Adobe has provided the ability to add form fields to static forms since Acrobat 3. In Acrobat 7 Professional, Adobe has finally made it easy to design smart forms and has even gone one step further and made it easy to compile the data in a spreadsheet or Access database.

Adobe LiveCycle Designer, previously a program purchased separately for approximately $1700 (US), is now packaged with Acrobat 7 Professional and makes creating forms a breeze. We are not going to teach the old Forms tools found on the Advanced Editing Toolbar: The Forms tools will be easy to use if you understand how to work with Adobe LiveCycle Designer.

In this Chapter

Follow along as we design the form – the chapter is designed as a single project. Therefore, there are no separate exercises or projects at the conclusion of the chapter. In this chapter you will use Adobe Designer to design a Personal Survey form. We will:

- Use New Form Assistant to get started
- Learn about the different form fields
- Learn how to construct the form
- Learn how to organize and name form fields
- Learn how to manipulate incoming data.

Note: In the chapter the program is referred to as Adobe Designer or Designer for simplicity.

Getting Started

The best way to learn is by doing, not just reading. Before you start, take a look at the finished form for reference. You can open Adobe Designer either from the Start menu or from within Acrobat.

Follow these steps to get started from Acrobat 7 Professional:

1. Click on the Forms Tasks button to open its menu (Figure 11.1). Select Create New Form to open an information dialog box explaining you are going to open a new program. Check "Do not show again" to disable the information dialog box that explains you are opening a separate program from within Acrobat. Click OK to dismiss the dialog box and open Adobe Designer.

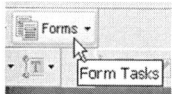

Figure 11.1 Use the commands on the task button menu

2. The program opens with the New Form Assistant displayed (Figure 11.2). At this point you can choose to design your own form or choose a template or existing document to use for the form and modify it for your own needs. For our project, click New Blank Form. Click Next to display the Setup options.

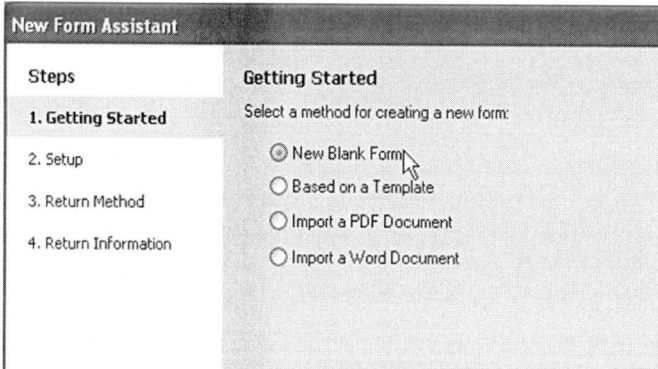

Figure 11.2 Use the New Forms Assistant for ease of design

3. On the Setup page, leave the default options for the page setup and number of pages and click Next.

4. Choose one of several ways to have the form returned to you on the Return Method screen (Figure 11.3). It is important to remember at this stage that only users with Acrobat 7 Professional or Standard will be able to save the form as they see it. Those with Adobe Reader will only be able to save the data file, not the actual form, or they can print the form. Select Fill then Submit/Print for this demonstration; click Next.

5. Use the fields on the Return Information screen if the form's data will be returned electronically. Type the recipient's email address and click Finish.

6. The wizard closes and the blank form opens in Adobe Designer.

Practice Files

This project is designed from scratch. For reference you can download and print the **Interest Survey Image for Exercise.pdf** file from the **ch11 folder** on the book's Web site to show you the field placement on the form page. The finished form, named **Interest Survey.pdf** is also available from the Web site in the **ch11 folder**.

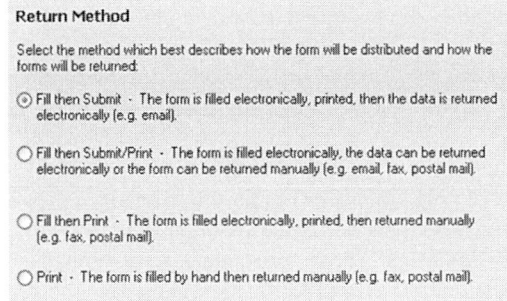

Figure 11.3 Define how the form contents are returned

Touring the Program

Next we will design our form, but first let us get a lay of the land and find out where all the tools are. The interface is shown in Figure 11.4.

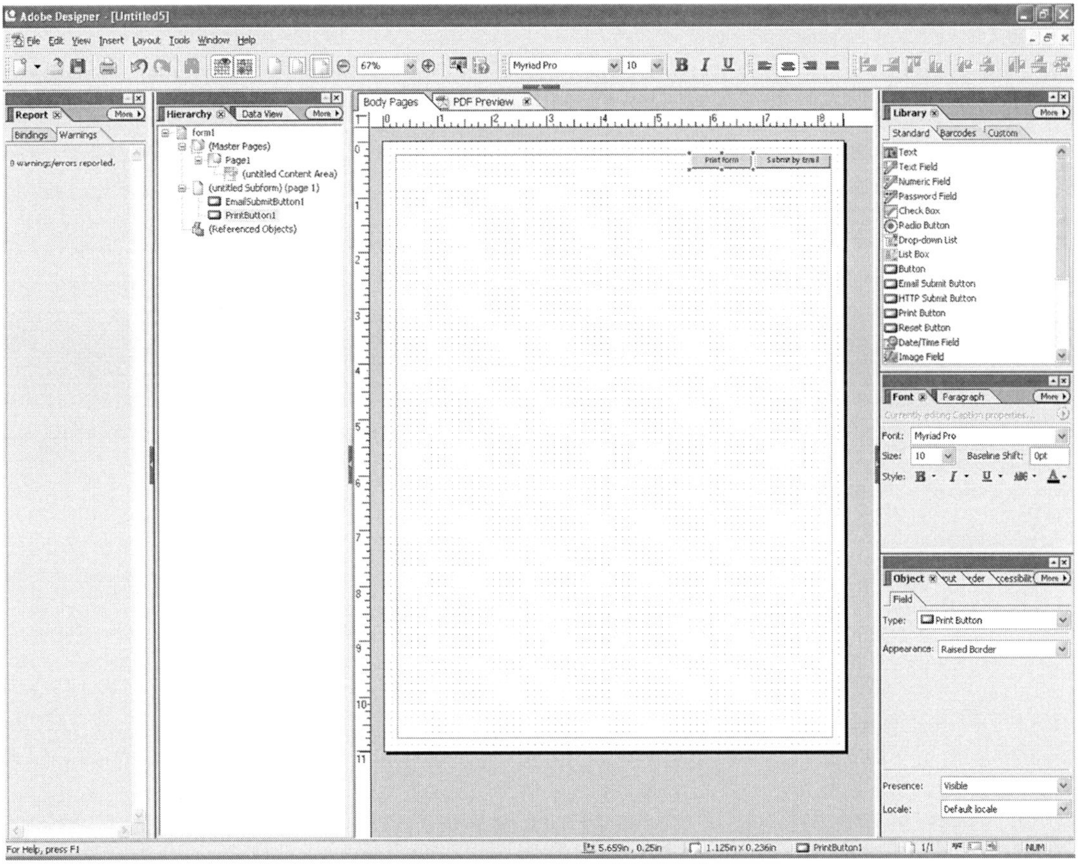

Figure 11.4 Adobe Designer's interface

- **Menu bar.** Like most programs, the menu bar stretches across the top of the program, and contains most of the program's commands. Including displaying/hiding palettes and alignment commands, useful for the design process.
- **Toolbars.** Below the menu bar are several commonly used toolbars; the active buttons depend on what part of the form is active.
- **Report palette.** This palette, on the far left side of the program window, lists form fields according to their binding type, as well as listings of any errors that may occur in the form. Show/hide the palette by clicking the blue arrow on the border of the palette about halfway down the window.
- **Hierarchy and Data View palettes.** These two palettes are tabbed together, and show the fields associated with the form. Clicking an object in the Hierarchy tab highlights the object on the form, thus allowing you to find content quickly in a long or complicate form. The Data View tab shows data connections for the form's fields. Show/hide the palette by clicking the blue arrow on the border of the palette about halfway down the window.
- **Body Pages.** The Body Pages view is the main working space used to design your form. The rulers and grid lines are shown by default, and can be toggled on and off in the View menu.
- **PDF Preview.** This nifty tool is tabbed with the Body Pages view. Select the PDF Preview tab to see how the form would look when viewed in Acrobat without having to save the file and open it in Acrobat. The preview is a great way to test your fields before publishing your form.

Palettes

Most of the program's tools are in palettes at the right of the program window, or are available from the Windows menu:

- **Library Palette.** The Library palette is composed of three tabs (Figure 11.5). Adobe has consolidated the standard objects in the Library palette and made it extremely easy to use. Under the Standard tab, Adobe has included many commonly used standard objects. Designer can support two types of barcodes, and these are managed on the Barcodes tab. To make a form designer's life easier, Adobe Designer allows you to create custom objects and store them under the Custom Tab instead of having to recreate a button each time you use it.

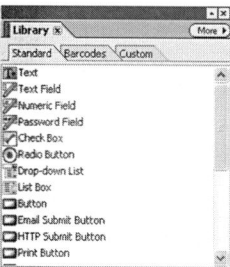

Figure 11.5 Choose standard objects from the Library palette

- **Text Palettes.** Below the Library palette is the Font and Paragraph palette. These tools are also available from a toolbar, but the palette is helpful when editing a field box with both a Caption and a Value. If a field is selected on the form, click on the Currently editing Caption and Value drop-down arrow to select an option (Figure 11.6). You can edit the Caption text, the Value text or both.
- **Objects and Other Palettes.** The bottom right of the program window includes a set of palettes used to modify object characteristics. The Objects palette includes fields for customizing objects you add to the form; adjust the form fields' positions and relationships on the Layout palette. Add frames to cells,

fields and subforms in the Border palette; add accessibility features like alternate text in the Accessibility palette.

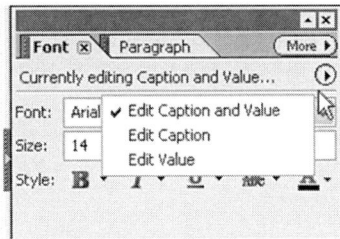

Figure 11.6 Edit captions, values, or both

Adding Form Content

Let us get back to the form design. Adobe Designer placed two buttons on our form because we chose to have the form either emailed to us or printed in the wizard during set up. We will move the buttons and then add more content, following these steps:

1. Shift-click to select both buttons and drag them to the bottom of the page. With the buttons still selected, choose Layout > Center in page > Horizontally.

2. Next add a title to the form. Click the Standard tab in the Library palette to display it. Click Text and drag to the form (Figure 11.7). A default Text block is added to the form.

3. Highlight the default text in the text block and type **Interest Survey**. Use the Font palette to increase the size of the font and center the title horizontally on the page.

4. Add another Text block and type a short explanation of the form's purpose. You can resize the text box by dragging its resize handles on the corners, or type dimensions in the Layout palette.

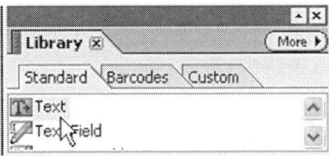

Figure 11.7 Drag an object to the Body Pages view to add it to the form.

Adding a Date Field

The first active field we'll add to the form is the date field. It is VERY important to name all your fields appropriately when creating the form if you are going to be collecting data electronically. If you choose names carefully now, creating/adding to your database later will be infinitely easier.

Follow these steps to add and name the first field:

1. Select the Date/Time Field object in the Standard tab of the Library palette, and drag a field to the left side of the form.

2. Select the default text and type **Today's Date**.

3. Click the Appearance drop-down arrow on the Object palette's Field tab and select Solid Box to change the field's appearance (Figure 11.8).

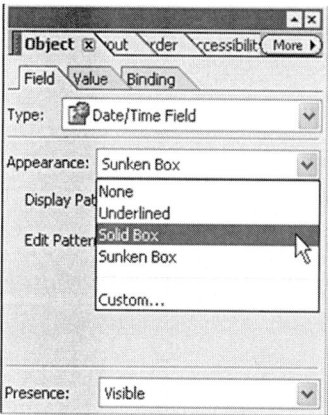

Figure 11.8 Change the field's appearance

4. Click the Display Pattern's drop-down arrow on the Field tab to choose a pattern or type a custom one as we are doing using a MM-DD-YYYY pattern.

5. In the Object palette, under the Binding tab, type **Date** in the Name field. Next to Data Pattern, type **MM-DD-YYYY**, which is how the data will be transferred back to you.

Specifying Required Fields

To define a field as "required" use the Object palette. Click the Value tab, and select the type to User Entered - Required type. If a user tries to submit the form without completing the required field, Acrobat displays an error message and highlights the field(s) that needs to be completed.

Adding a Text Field

Next it is time to add fields for the user's information. Follow these steps to add and configure the first field:

1. Click the Text Field object on the Standard tab of the Library palette and drag a text field to the form. Change the caption to First Name.

2. On the Field tab in the Object palette, click the Appearance drop-down arrow and choose Underlined. The sunken box turns into a line.

3. Click the Layout palette, and the click the Position drop-down arrow in the Caption section and choose Bottom; the caption is now centered under the field's line (Figure 11.9).

4. Choose Align Center in the Paragraph toolbar; and change the font size to 12pt in the Font toolbar (Figure 11.10).

Note: Designer will not allow spaces in the Name field as data connections do not allow for blank spaces. Decide on the naming style you are going to use before starting the project, such as capping each element in a hierarchical name or using underscores.

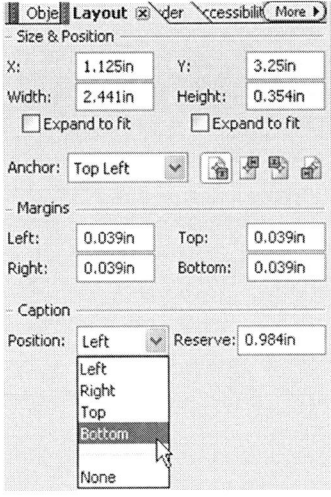

Figure 11.9 Use the tab's options to configure the field

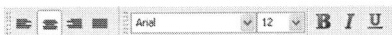

Figure 11.10 Modify characteristics in the toolbars

5. The caption disappears when you reset the size, and a red (+) shows on the field's frame, indicating that the field is too small to hold the text. Adjust the position of the line by clicking on the orange line and drag it up until the caption is visible (Figure 11.11).

6. Size the field in the Layout Palette. Change the Width to 2.25in and the Height to 0.75in. Click Expand to fit for the dimensions to allow the form to change size to fit entered data.

7. Change the X position to 1.0in. to move the field to the left of the form.

8. Finally, type **FirstName** in the Name field on the Object palette's Binding tab.

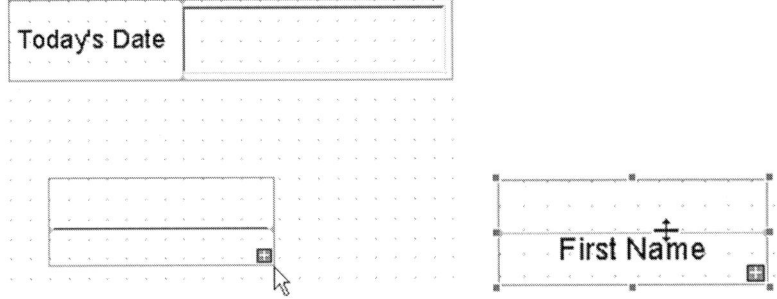

Figure 11.11 An indicator displays if a text box is too small (left); resize the field to show its contents (right)

Creating a Custom Field

The LastName field needs to have the same attributes as FirstName. One could go through the process again and create the field from scratch, but since we will likely need this field again, let us create a Custom field.

Follow these steps to save a custom field:

1. Click the Custom Tab in the Library palette. You will see that Adobe Designer has already populated this list with some commonly used fields (Figure 11.12).

2. Click on the First Name text field and drag it to the Custom tab. You will notice there is now a plus sign under your arrow.

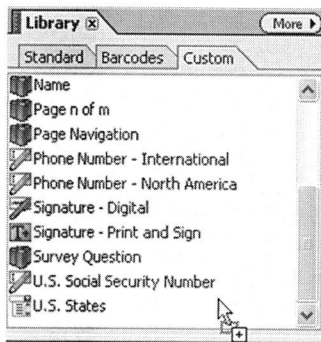

Figure 11.12 Adobe Designer provides default fields for customizing

3. Release the mouse and the Add Library Object dialog box opens. Name the field and add a brief description. You can also choose under which tab the field is to be placed. Name the object **User Name**.

4. Click the User Name object and drag it to the form, aligned to the right of the **FirstName** field.

5. Change the Caption to read **Last Name** and change the Name of the field in the Binding tab to **LastName**.

6. Shift-click both text fields and then choose Layout > Align > Top to align the two fields.

Adding Selection Fields

The types of field you use are based in part on how convenient it is for the user to answer, as well as whether you want one or more answers.

The next items on our form are different questions the user answers by selection:

- The first question asks for the user's gender, which our users will answer by selecting a radio button.
- Then we will add two drop-down lists of items from which the user can select one choice in answer to each survey question.
- Finally, we will add checkboxes, from which the user can select more than one answer.

Adding Radio Buttons

We will add a radio button to force the user to make an either/or choice. Follow these steps to add the radio button:

1. From the Library Palette, choose Radio Button and drag it to the form next to **Last Name**; align the tops of the two fields.

2. Change the caption for the radio button field to read **Male** and change the font size to 12pt in the Font toolbar.

3. Name the field **Gender** in the Binding tab; you will see that the Value 1 now means Male (Figure 11.13).

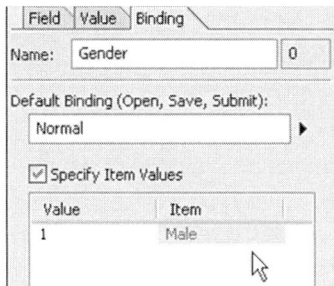

Figure 11.13 The first radio button value is male

4. We also need a Female choice. Right-click on the Male radio button to open the shortcut menu and choose Copy, then right-click below the Male radio button and choose Paste from the shortcut menu.

5. Using the Align commands in the Layout menu, align the two buttons to the left, and align the lower radio button with the bottom of the Last Name text field.

6. Change the button field's caption to **Female** in the Name field in the Binding tab. You see the Value 2 listed as Female (Figure 11.14).

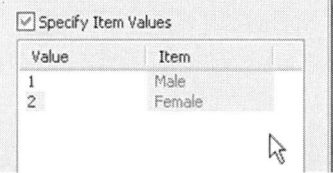

Figure 11.14 The second radio button value is female

7. Select the PDF Preview (tabbed with the Body Pages view) and test the form. Notice you must toggle between the two choices but cannot choose both (Figure 11.15).

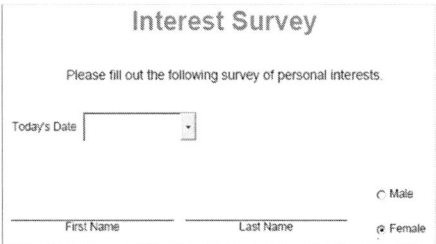

Figure 11.15 The radio button lets you make only one choice

Using a Drop-down List

In the Acrobat Forms tools, a drop-down list is referred to as a *combo box*, meaning a list of options with assigned values from which the user makes a choice. The form's creator can define the list or allow the user to input data. We are going to create two drop-down lists: one for favorite color and one for favorite snack.
   Follow these steps to add the drop-down lists to the form:

1. Select the Drop-down List object from the Standard tab on the Library palette, and drag it to the form. Place it below the First Name text field.

2. Change the caption to **Favorite Color**; and change the font size to 12pt. You will have to resize the box to read the caption after changing the font size. Name the field **Color**.

3. Click the Field tab in the Object palette; you see the Drop-down List is selected as the Type. Click the green (+) to activate the List Items text field and type the first item. Add a sequence of items, either press Enter or click the green (+) to add additional items.

4. For the first drop-down list's items, type: Pink, Purple, Red, Yellow, Green, Blue, Black, White.

5. Order the list alphabetically by clicking a value and using the arrow keys to move it up or down the list (Figure 11.16).

6. For the finishing touches, name the box **Color**, set the width to 3inches, the height to .75inches, and change the appearance to a Solid Box.

7. Repeat the same process to create another drop-down list of favorite snacks. Name the field Snack. Use the caption **Favorite Snack**.

8. Use these items for the list: Cake, Candy, Chips, Fruit, Veggies.

9. Arrange the two drop-down lists side by side on the form. Your form should now look similar to Figure 11.17.

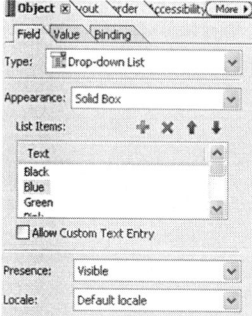

Figure 11.16 Add values and set appearance options for the drop-down list

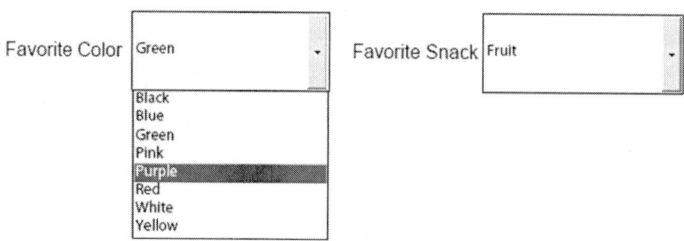

Figure 11.17 Add and configure two drop-down lists

Placing Check Boxes

Check Boxes allow the user to choose one or more items from a variety of choices. We are going to create a series of 12 check boxes to inquire about the user's hobbies.

Rather than building each field separately, we are going to build and configure the first field, and then quickly create multiples.

Follow these steps to add survey questions:

1. Add a Text object from the Standard tab of the Library palette to the form and type instructions for the questions. Our form says **Hobbies: please check all that apply**. Change the font size to 12pt.

2. Place the Text box below the Favorite Color question and left align the text on the form.

3. Next, from the Library palette, drag a Check Box object from the Standard tab to the form below the Hobbies text.

4. Change the font size to 12pt and the width to 1.5in. Change the appearance of the square to a Solid Square. Move the caption segment of the field so it is next to the check box.

5. Select the check box field and choose Edit > Copy Multiple to open the Copy Multiple dialog box (Figure 11.18). We will make a row and then duplicate the row.

6. On the dialog box, type 3 in the Number of copies field, and click the No vertical movement radio button (Figure 11.18). Click the Place to the right radio button; click the Offset by radio button and type 1.5in in the field, which represents the distance between the copied fields on the form. Click OK to close the dialog box and add the additional fields (Figure 11.19).

Tip: If you try to make vertical and horizontal placement at one time, Adobe Designer will place them in a diagonal fashion on the form and place only the number of copies specified in the Number of copies box.

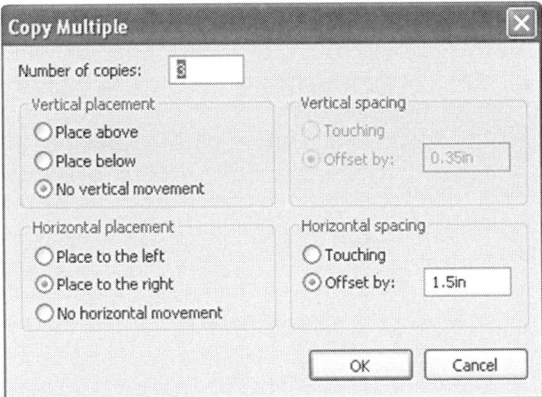

Figure 11.18 Set the values for the first row.

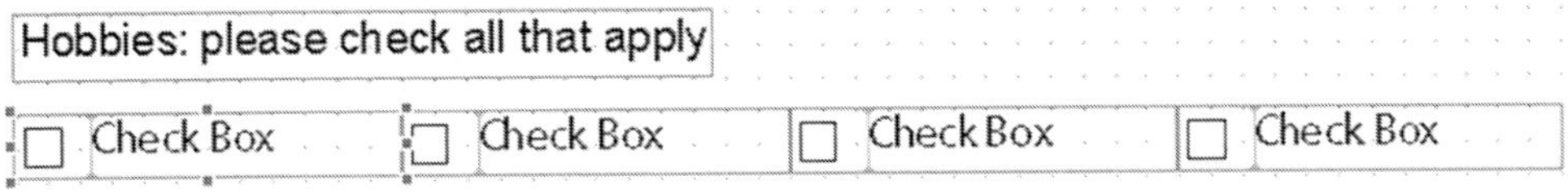

Figure 11.19 The first row has four checkbox fields

7. Shift-click the set of four check boxes, and choose Edit > Copy Multiple again to reopen the dialog box. This time, choose to make two copies, click the Place below vertical placement radio button; click the No horizontal movement radio button. Click the Offset by radio button and type .25in and click OK to close the dialog box and add the additional check boxes.

Finishing up the Checkbox Fields

The list of options for the hobbies we use are: Reading, Hunting, Sleep, Exercise, Sports, Computer, Games, Horses, Television, Art, Food, Nothing. The finished check boxes are shown in Figure 11.20.

Select the default text for each check box and replace it with one of the options. Rename each check box in the Binding tab of the Object palette to reflect its appropriate name. For example, the Reading check box would be named **HobbyReading**, the Exercise checkbox would be **HobbyExercise**, and so on.

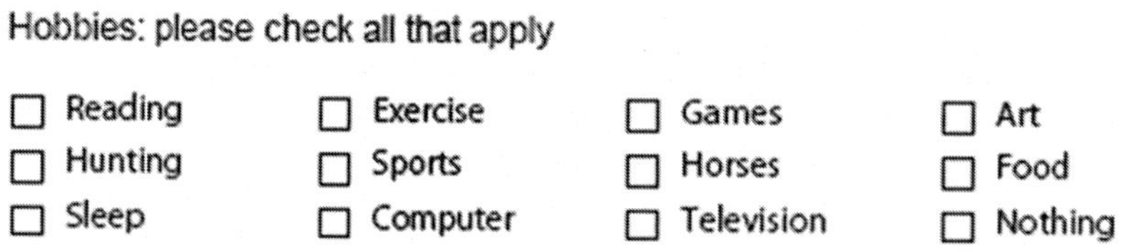

Hobbies: please check all that apply

☐ Reading   ☐ Exercise   ☐ Games   ☐ Art
☐ Hunting   ☐ Sports   ☐ Horses   ☐ Food
☐ Sleep   ☐ Computer   ☐ Television   ☐ Nothing

Figure 11.20 The set of check boxes are evenly distributed and spaced

Using Numeric Fields

A numeric field is similar to a text field, in that the user manually enters the information, yet a numeric field only allows numbers to be entered. Our form uses two numeric fields: Number of Pets and Number of Cars.

Follow these steps to add and configure numeric fields:

1. Select the Numeric Field object in the Standard tab of the Library palette and drag it to the form, aligned to the left below the Hobbies question.

2. Set these options for the field:
   • Caption: Number of Pets
   • Font size: 12
   • Appearance: Underlined
   • Name: Pets

3. Copy the field and place it to the right of the Pets field.

4. Set these options for the second numeric field:
   • Caption: Number of Cars
   • Font size: 12
   • Appearance: Underlined
   • Name: Cars

5. Adjust the fields so you can read the captions and they are centered under the Hobbies section. The finished fields are shown in Figure 11.21.

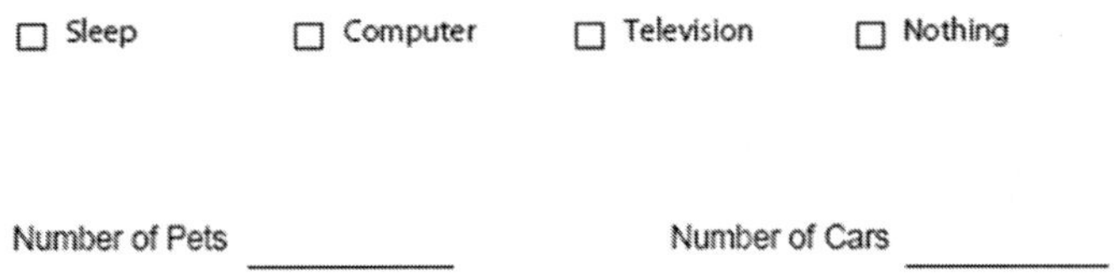

☐ Sleep   ☐ Computer   ☐ Television   ☐ Nothing

Number of Pets   _____   Number of Cars   _____

Figure 11.21 The form includes two numeric fields

Adding and Modifying Buttons

Buttons are objects on the form the user pushes to execute a command. The form originally includes two buttons based on our selected method of return. Most users also like to have a Reset button on the page, so they can restart the form again without having to close and reopen the file. Adobe Designer includes such a preconfigured option.

To add a button to the form, select the Reset Button object in the Standard tab of the Library palette and drag it to the upper right corner of the form. Since the button is a defined object, Adobe Designer includes the code for the button. If you want to create a button which executes a different command, drag the generic Button on the form and you can attach the appropriate script to the button.

There are three buttons on our form. Let us change the two buttons' appearance at the bottom of the form. Shift-click to select both the Print Form and Submit by Email buttons. Change their font size to 14pt, the width to 2in, and the height to .75in. Center both buttons horizontally on the page.

Finishing Touches

The form looks presentable as it is, but is rather boring. We are going to add some boxes around the sections to break it up visually by following these steps:

1. Select the Rectangle object in the Standard tab of the Library palette.

2. Click your mouse on the form at the upper left of the First Name section. You will notice the pointer is now crosshairs with a rectangle to the lower right of it, which signifies you are about to draw a rectangle (Figure 11.22).

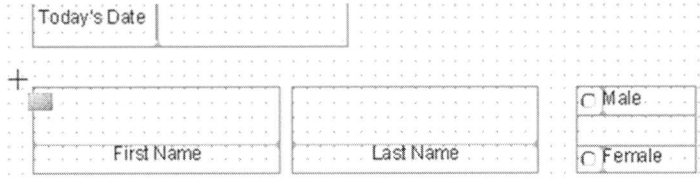

Figure 11.22 The cursor identifies the object being drawn

3. Drag to enclose the section with a rectangle. Release the mouse to complete the rectangle.

4. Repeat with the remaining sections of the form, for a total of four rectangles (Figure 11.23).

5. To organize the rectangles' appearance, shift-click to select all the rectangles and then choose Layout > Make Same Size > Width. Repeat choosing Layout > Align > Horizontal Center.

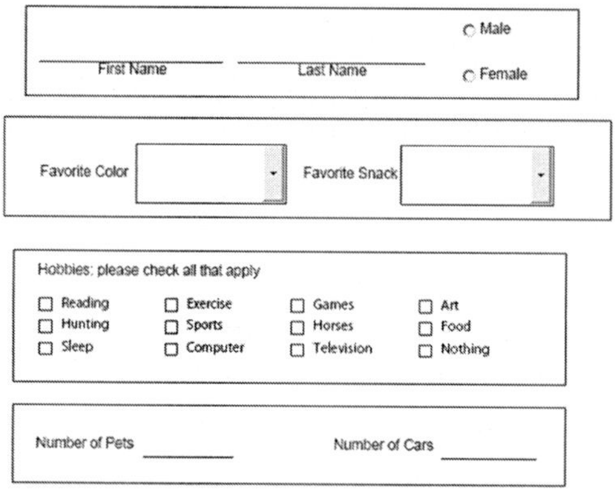

Figure 11.23 Add four rectangles to the form.

Great job!! The form is complete, and looks like the image you printed out at the start of the project (Figure 11.24).

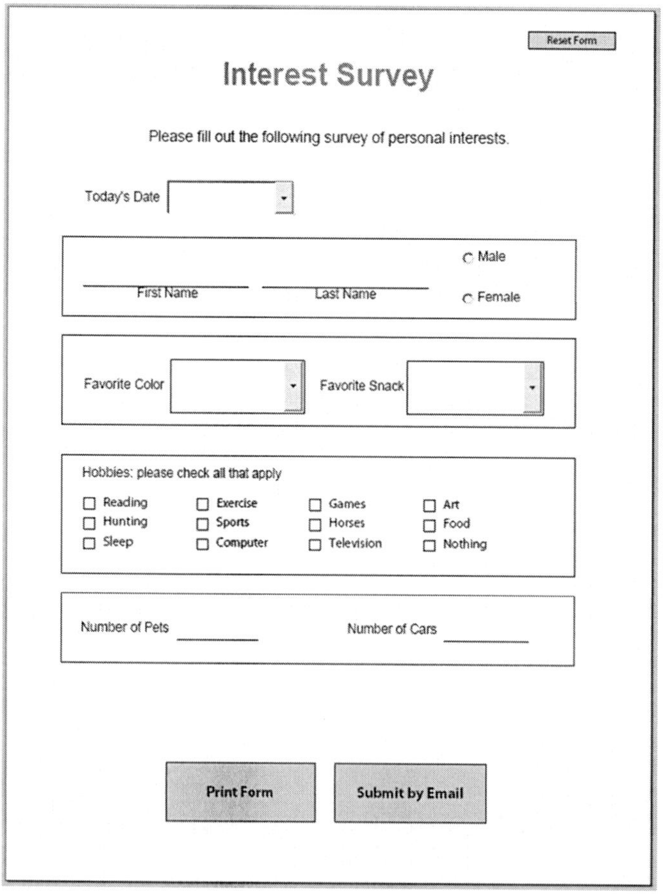

Figure 11.24 The completed survey form

Collecting Data

Hooray! We have completed the form and sent it out to all our friends. Now our email box is full of messages with an attachment named "Data form Survey1". Now what do we do?

If you open an attachment, you see it is a collection of data, which is not very helpful, although you can easily see how the user has answered the survey questions (Figure 11.25).

Acrobat 7 Professional lets you import the information from each return individually into the original form. You can also send the data files to an Excel spreadsheet which then can be converted to a database or manipulated in many fashions, which we are going to do.

First, save all your data files in a folder, because you cannot directly open the data files in Acrobat from your email. Be careful while saving: when the data files are returned to you they are usually named the same. Rename each file according to a preferred naming schedule. You should open the file and then save it rather than saving the file directly from the email to save the data files in XML format (Figure 11.26).

Project Files

A set of data files for the form is available in the **ch11 folder** on the book's Web site.

```
-->
<DateTimeField1>2005-04-15</DateTimeField1>
<FirstName>Ima</FirstName>
<LastName>Wright</LastName>
<Gender>2</Gender>
<Color>Green</Color>
<DropDownList1>Chips</DropDownList1>
<HobbyReading>0</HobbyReading>
<HobbyExercise>1</HobbyExercise>
<HobbyGames>0</HobbyGames>
<HobbyArt>0</HobbyArt>
<HobbyHunting>0</HobbyHunting>
<HobbySports>0</HobbySports>
<HobbyHorses>0</HobbyHorses>
<HobbyFood>1</HobbyFood>
<HobbySleep>0</HobbySleep>
<HobbyComputer>0</HobbyComputer>
<HobbyTelevision>0</HobbyTelevision>
<HobbyNothing>0</HobbyNothing>
<Pets>10.00000000</Pets>
<Cars>1.00000000</Cars>
</form1>
```

Figure 11.25 Returns are strings of data

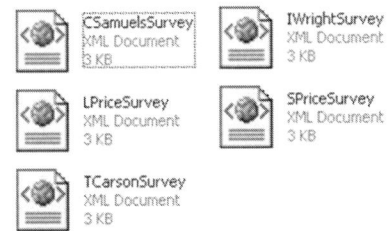

Figure 11.26 Save the returned data files

Data to Form

First, we want to look at everyone's data in the actual form; open the completed survey form in Acrobat 7. To import survey results, follow these steps:

1. Choose File > Form Data > Import Data to Form (Figure 11.27).
2. The Select File Containing Form Data dialog box opens.
3. Locate and choose the file you would like to load and click Select.
4. The dialog box closes, and the blank form is populated with the data from the imported file.

You can save this form with the information stored if you want.

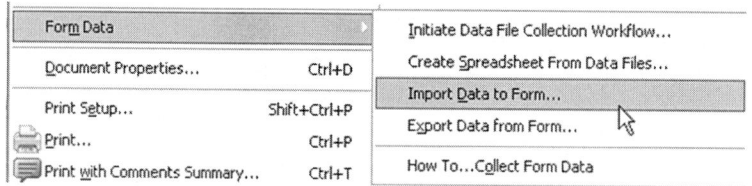

Figure 11.27 Acrobat 7 includes several form data commands

Data to Spreadsheet

Next, we are going to export the results from all our returned forms to a single Excel spreadsheet. Acrobat 7 has a handy feature that automatically builds a spreadsheet.

Follow these steps:

1. Choose File > Form Data > Create Spreadsheet From Data Files to open the Export Data From Multiple Forms dialog box.
2. Click Add Files; in the resulting dialog box, locate and select the files you want to add to the spreadsheet and click Select to list the files in the export dialog box (Figure 11.28).
3. Click Export to dismiss the dialog box and open the Select Folder to Save Comma Separated File dialog box. Choose a name and location for the file and click Save.

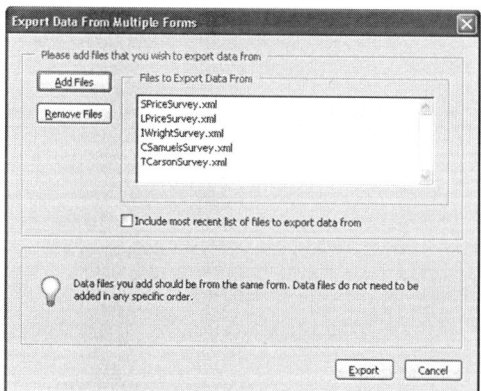

Figure 11.28 Select the files to export to the spreadsheet

4. The Export Progess dialog box displays, showing the processing of the files (Figure 11.29). When complete, click either View File Now or Close Dialog.

5. Click View File Now to open the Excel spreadsheet (Figure 11.30).

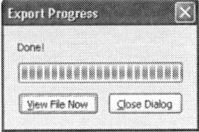

Figure 11.29 The selected data files are processed

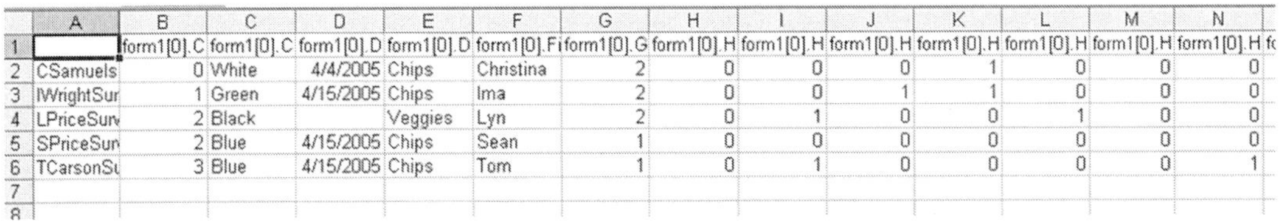

A	B	C	D	E	F	G	H	I	J	K	L	M	N
	form1[0].C	form1[0].C	form1[0].D	form1[0].D	form1[0].Fi	form1[0].G	form1[0].H	form1[0].H	form1[0].H	form1[0].H	form1[0].H	form1[0].H	form1[0].H fc
CSamuels	0	White	4/4/2005	Chips	Christina	2	0	0	0	1	0	0	0
IWrightSur	1	Green	4/15/2005	Chips	Ima	2	0	0	1	1	0	0	0
LPriceSur	2	Black		Veggies	Lyn	2	0	1	0	0	1	0	0
SPriceSur	2	Blue	4/15/2005	Chips	Sean	1	0	0	0	0	0	0	0
TCarsonSu	3	Blue	4/15/2005	Chips	Tom	1	0	1	0	0	0	0	1

Figure 11.30 Form data are added to a spreadsheet

The field names on the spreadsheet are prefaced with form1[0]. You may need to manually rename the fields to the appropriate names if exporting to a current database.

Summary

In this chapter you were introduced to Adobe's Form Designer, a full-featured program integrated into Acrobat 7 Professional for Windows. The chapter walked through the development of a simple survey form from a blank form, and showed you how to add different types of field.

You saw how to name fields, and how to bind data to a field. You also learned how to incorporate data files into a source form, as well as a method of exporting form data to a spreadsheet for examination and evaluation.

12

Commenting and Reviewing

One of Acrobat's most powerful features is its ability to display comments on documents that you can add using a wide variety of tools. Not only does Acrobat provide you with multiple toolbars full of options, but you can customize and set the default appearance of any tool to meet your needs.

Acrobat 7's ability to enable documents on a file-by-file basis for commenting in Adobe Reader 7 is revolutionary – effectively extending the program's reach to millions more users.

The Review Tracker introduced in Acrobat 6 has changed its name to the Tracker in Acrobat 7 and become a separate window from the rest of the program. From this one interface you can initiate, manage, and participate in reviews, take files offline, and work with associated entities such as RSS feeds and newsreaders.

In this Chapter

Acrobat offers tools and processes to perform communication cycles among two or more review participants.

In this chapter you will discover how to work with the range of commenting tools offered in Acrobat and how to use a review cycle:

- Commenting tools are split into several toolbars, each designed for different general uses.
- Any comment can be customized when applied to a file, and default appearances specified.
- Acrobat contains a number of specialized viewing tools, often used in conjunction with commenting and markup tools in AEC projects.
- Comments can be manipulated on the document page, or within the Comments pane.
- Reviews can be initiated using a simple wizard, and tracked and managed through the Tracker.

Commenting Tools

There are several toolbars and subtoolbars in Acrobat containing a wide variety of different types of tool used for commenting and document markup. The best way for you to understand how the tools work is to experiment. You may find that certain tools become invaluable in your particular project workflows. Acrobat 7's commenting tools can be accessed from the Comment & Markup task button's menu, the Commenting menu, from the Comments pane, and the Commenting toolbar.

Table 12.1 breaks the Commenting toolbar into sections and explains each tool's purpose.

Table 12.1 Commenting toolbar

Icon	Type of tool...
	Note tool adds collapsible note boxes to a document
	Text Edit tools let you indicate text edits on a document; contains a subtoolbar of editing options
	Stamp tool used to place a stamp on a document, Includes a menu of stamp types, management dialog boxes, and customizing options
	Highlighting tools are electronic versions of traditional highlighter, cross outs, and text underlines
	Attach either a file or a voice comment using these tools
	The items on the Show menu are used to access, sort, and view contents of comments

Managing Properties

Experiment with the Properties bar rather than using comments' Properties dialog box. You may find you prefer its convenience. Right-click the toolbar area at the top of the program window and choose Properties Bar. The content of the toolbar changes depending on the object selected or active tool.

Customizing a Comment's Appearance

Every comment added to a file can be customized to a greater or lesser extent, depending on the type. All comments can minimally have their colors changed; any comment that includes an icon, such as a note, can be customized further.

Follow these steps to customize a Note comment's appearance:

1. Click the Options button on the Note box or right-click the Note box or the comment icon to display the Options menu and click Properties to open the Properties dialog box, showing the Appearance tab.

2. Select a different icon for the comment from the list on the Appearance tab (Figure 12.1).

3. Click the color swatch and choose a color for the comment and its icon from the color palette, also shown in Figure 12.1.

4. Select the General tab and change the name of the commenter; you can also change the subject and lock the comment to prevent changes.

5. Click Close to close the dialog box.

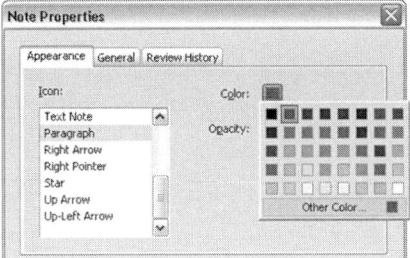

Figure 12.1 Choose visual settings for the comment

Tip: If you want a custom color, click Other Color on the color palette to open a Color Picker; adjust the opacity of the comment by dragging the Opacity slider, hidden beneath the color palette in Figure 12.1.

Making a Text Edit

Acrobat and Adobe Reader include a set of tools used for identifying text edits. The Text Edits tools are used to identify content to replace, delete, or insert. Follow these steps to select text for editing:

1. Click the Text Edits tool on the Commenting toolbar to open the submenu and select the Indicate Text Edits Tool .

2. Click and drag with the tool to select text on the page; change the amount of text selected by dragging one of the arrows at the corners of the highlighted text (Figure 12.2).

3. Select the appropriate tool or keyboard command and make the edit:

   - Press Delete on the keyboard to delete the selected text, shown as a cross-out on the page, or select the Cross Out Text for Deletion ⊤ comment tool.

   - Start typing to make a Replace Selected Text comment, or choose the tool ⊤ₐ from the Text Edits submenu, shown as a cross-out on the page with a caret ᵘᵗₐ identifying the location at which the replacement text starts. The replacement text is shown in a note box automatically.

   - Rather than selecting text, click the location in a block of text with the Indicate Text Edits tool and start typing to insert text; or choose the tool from the Text Edits Tₐ submenu. The text is shown in a note box and a caret displays on the page at the insertion point.

4. Double-click the edit to open a note box to type a comment or note.

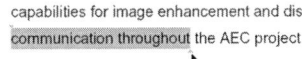

Figure 12.2 Select the text for editing

Adding Stamp Comments

Acrobat offers you the digital equivalent to old-fashioned ink stamps, and takes the features much further by using dynamic features such as names and dates for some stamps. Select from a variety of preconfigured stamps, or create your own custom stamps.

Apply stamps as you would other comments:

1. Click the Stamp icon on the Commenting toolbar to open its menu, and then choose a stamp from the submenus (Figure 12.3).

2. Move the cursor over the page, which displays a semi-transparent version of the stamp, and click to place the stamp.

3. Double-click the stamp on the page to open a note box to contain explanatory text.

Creating a Custom Stamp

You can simply create custom stamps to use in your reviewing and markup tasks. You might want a stamp showing your company's logo, for example, or a text stamp such as "Approval Pending".

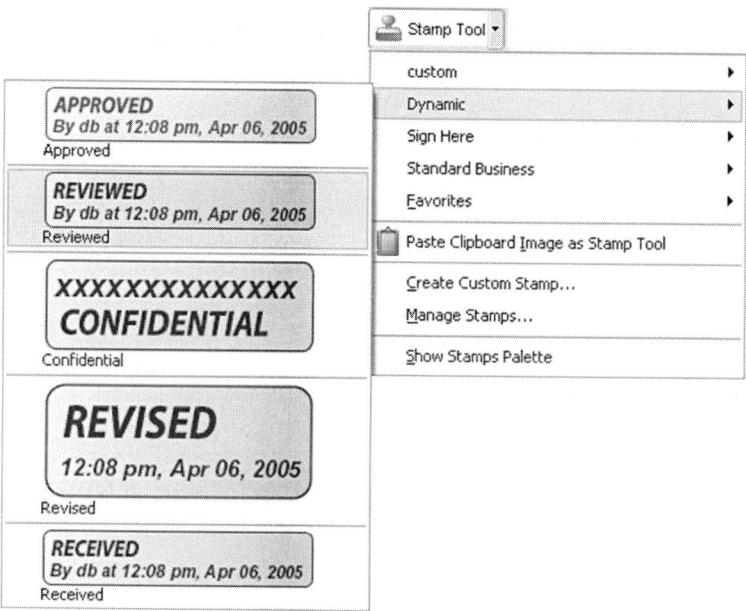

Figure 12.3 Select a preconfigured stamp

Follow these steps to create a custom stamp:

1. Click the Stamp Tool's drop-down arrow on the Commenting toolbar to open its menu and choose Create Custom Stamp to open the Select Image for Custom Stamp dialog box.

2. Click Browse to locate and select the file you want to use for the stamp. You see a sample on the dialog box (Figure 12.4).

3. Click OK to close the dialog box, which is replaced with the Create Custom Stamp dialog box (Figure 12.5).

4. Click the Category drop-down arrow and choose a category in which to store the stamp. You can also name a new category from this menu.

5. Type a name in the Name field and click OK to close the dialog box.

6. The new stamp is added to the specified category and ready for use.

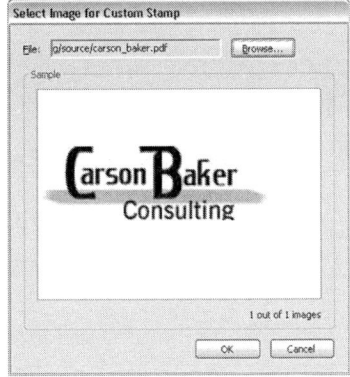

Figure 12.4 Locate the file to use for the stamp

Tip: Right-click a customized comment and choose Make Current Properties Default to change its appearance.

Other Types of Comment

You can use a file as a comment identified on the document by an icon which your user double-clicks to open in its native application. You can also use a recorded voice or other audio message as a comment.

Attaching a File or Voice Comment

Attaching a file as a comment also lists the file in the Attachments pane. Read about using attachments in Chapter 8. To attach a file as a comment:

1. Click the Attach a File as a Comment tool on the Commenting toolbar.
2. Click the page where you want the icon placed; the Add Attachment dialog box opens.
3. Locate and select the file you want to attach.
4. In the File Attachment Properties dialog box, make changes as desired or click Close to dismiss the dialog box.

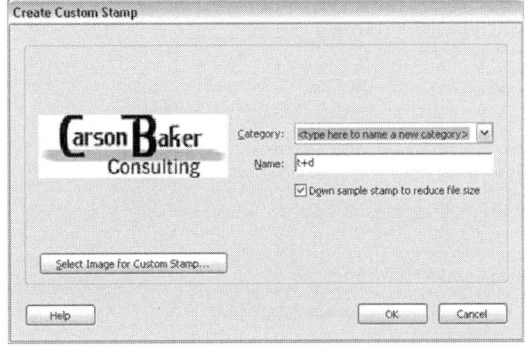

Figure 12.5 Name and categorize the new stamp

Recording and Using Audio Comments

Use the Attach Voice Comment tool if you prefer to record a voice comment rather than attaching a written comment to a document. The comment is embedded within the document, and is not listed as a separate attachment to the file. Acrobat supports AIFF and WAV file formats.

To use a voice comment:

1. Click the Record Audio Comment tool 🔊 on the Commenting toolbar to select it.
2. Click the page where you want to place the comment's icon; a Sound Recorder opens (Figure 12.6).

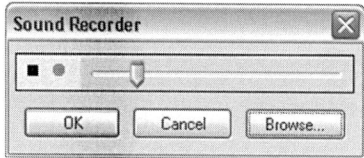

Figure 12.6 Record or locate the voice message

3. Record your message; or click Browse to locate and open an existing recording.
4. Click OK to close the Sound Recorder and open the Sound Attachment Properties dialog box.
5. Choose any customizations, such as the appearance of the icon and color 📎.
6. Click Close to dismiss the dialog box.
7. To play the comment, double-click its icon on the page, or right-click the icon and choose Play File from the shortcut menu.

Specialized Viewing Tools

Acrobat 7 offers several viewing tools often used in conjunction with examining and commenting on a document. Click the displayed Zoom tool on the Zoom toolbar to open the menu (Figure 12.7). You can also find the Zoom tools on the right-click shortcut menu. The tools include different types of zoom tool, pan and zoom, and "instant" magnification options:

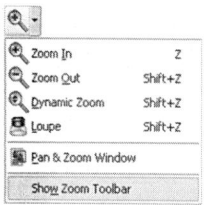

Figure 12.7 Use one of several tools for controlling the view

- Use the Zoom In 🔍 or Zoom Out 🔍 tools to increase or decrease the magnification: switch from one tool to another by pressing Ctrl as you click with the tool.
- Use the Dynamic Zoom tool 🔍 to move the magnification higher and lower without having to select different tools or use keystrokes.

- Use the Loupe tool 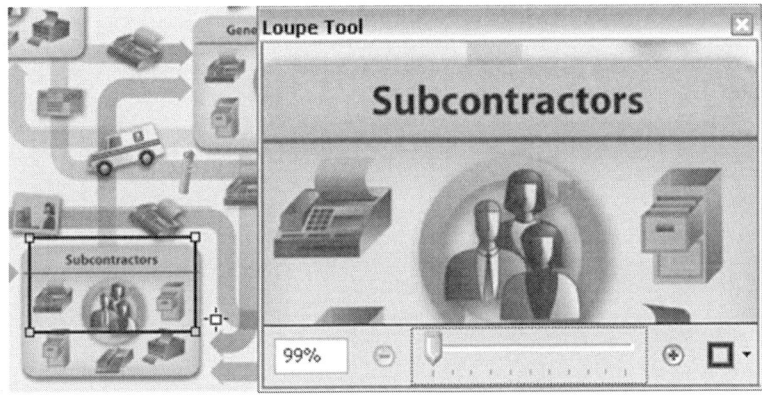 to magnify a specific area of the page in context. Click the document with the Loupe tool to activate a small window; this window shows the area identified by the rectangle on the Document pane (**Figure 12.8**). Drag the edges of the rectangle, use the (+) and (–) buttons, or drag the slider to change the content and magnification of the area displayed.

- Use the Pan & Zoom tool to scan a document at a high magnification. Choose the tool on the Zoom menu to open a secondary window over the main program window (**Figure 12.9**). Drag the Pan & Zoom box around the page to show magnified detail. Change magnification using the (+) and (–) buttons, manually typing a value, or resizing the rectangle overlay.

Figure 12.8 The Loupe tool magnifies a specified area

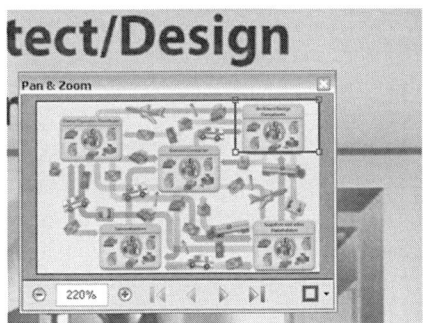

Figure 12.9 Use Pan and Zoom to view detail

Using Tools in Combination

Use both the Pan & Zoom window and the Loupe tool when you want to scan a document and check out magnified details at the same time.

Follow these steps to work with both tools:

1. Click the Pan & Zoom Window option from the Zoom tool drop-down menu.

2. When the Pan & Zoom Window opens, use the controls to move around the window.

3. Click the Loupe tool on the Zoom tool menu.

4. Click the window with the Loupe tool for a closeup view.

Note: In both Loupe and Pan & Zoom tools' windows, click the colored rectangle to open a Color Picker and select a different color to see the overlay on the document more clearly.

Initiating a Review

Although a review can be set up using Acrobat 7 Standard, you cannot provide additional rights to users working with Adobe Reader unless you enable the file in Acrobat 7 Professional. Any file can be attached to an email, of course, but use the wizard in Acrobat 7 to set up the review using the Tracker to control the document. Follow these steps to initiate a review:

1. Click the Send for Review task button's pull-down arrow and choose Send by Email for Review from the menu.

2. The three-paned Send by Email for Review wizard opens (Figure 12.10).

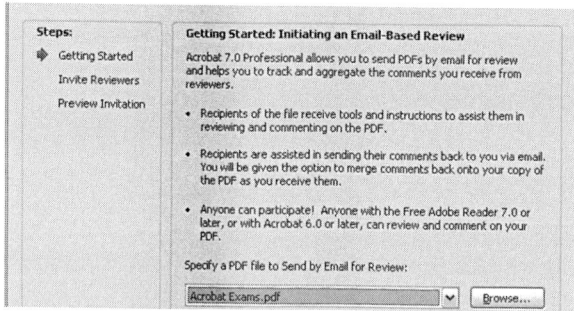

Figure 12.10 Select the command to start the review process

3. Click the drop-down arrow on the Step 1 pane and choose an open document, or click Browse and locate and select the document to use for the review in the Open dialog box.

4. Click Next to display the second pane of the dialog. List the names of the recipients in the Invite Reviewers field at the top of the dialog box (Figure 12.11). Click the Address Book button to load addresses from your Outlook address book.

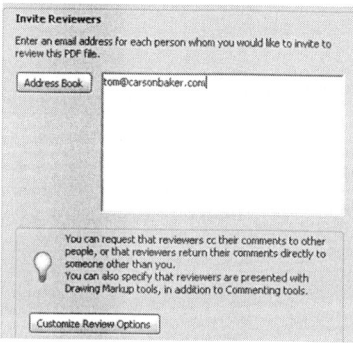

Figure 12.11 List addresses of review participants

5. To customize the options for your reviewers, click Customize Review Options to open the Review Options dialog box (Figure 12.12). Make selections as required, and click OK to return to the wizard. You can customize:

- The email address you want the comments sent to, if other than yours.
- Allow users to use Drawing Markups, which are a subset of commenting tools by clicking the check box.
- Allow users working with Adobe Reader 7 to use commenting tools.

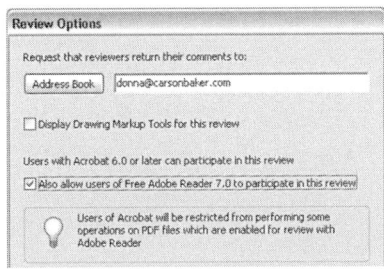

Figure 12.12 Customize the review if necessary

6. Click Next on the wizard to display the final pane showing a preview of the invitation. Customize the text in the subject and contents of the email if desired.

7. Click Send Invitation to transfer the message to your email program and attach the PDF file.

8. An Outgoing Message Notification information dialog opens explaining what happens next. Depending on your system configuration, you may see further dialog boxes; follow the prompts and dialog boxes to distribute the email.

Enabling Impacts

When you enable a document for reviewing in Adobe Reader 7, some functions are restricted in Acrobat 7, such as inserting and deleting pages, editing content, signing the document, and filling form fields.
    A document can be enabled using the Comments > Enable for Tracking in Adobe Reader command from the program menu. If you use this method, resave the file to include the feature in the document.

Managing a Review in the Tracker

Reviews that you have initiated or are participating in can be managed in Acrobat's Tracker. Choose Comments > Tracker, or click the Send for Review Task button and choose Tracker from its menu. The Tracker opens as a separate window (Figure 12.13).
    To work with reviews:

- Click the (+) to the left of My Reviews in the left column of the Tracker to open the list of active reviews.
- Click the name of a review you want to check on.
- Details of a selected review are shown in the right pane.
- The addresses of invitees to a review are hyperlinks; click a link to open an addressed email window.
- Manage the review using the options from the Manage menu.
- When you are finished with a review, click Remove to delete it from the Tracker.

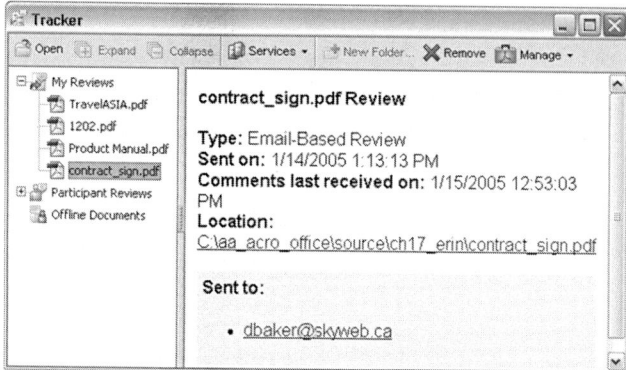

Figure 12.13 Control communication with review participants from the Tracker

Other Tracker Services

Use the Tracker as a news reader and for accessing broadcasting services. Locate the RSS (Really Simple Syndication) feed you want to subscribe to and copy its URL from the Web page.

In Acrobat's Tracker window, click the Services pull-down arrow to open a menu and choose Subscribe. An Add Subscription dialog box opens, and displays the URL you copied from the feed's Web page; click OK to close the dialog box and list the subscription in the Tracker by name. Click a subscription name to show a summary in the right column of the dialog; click the summary to display the full text in your browser.

Conducting a Browser-based Review

A browser-based review lets you access files at will, as long as the files are uploaded to a server folder accessible by Internet. Browser-based reviews can be used in Windows, and are supported on the Mac using Safari 1.3, and Mac OS 10.3.x.

Follow these steps to initiate a browser-based review:

1. Choose Upload for Browser-Based Review from the Send for Review task button's drop-down menu.

2. Follow the prompts in the wizard, which is very similar to the Send by Email for Review wizard.

3. Select a server folder in the wizard. Choose and configure a server for reviewing either in Acrobat's preferences or on an individual file basis.

4. A browser-based review is listed in the Tracker.

Specifying a Server

You can either specify a server configuration on a file-by-file basis, or set a server in Acrobat's Preferences by following these steps:

1. Choose Edit > Preferences > Reviewing.

2. Click the Server Type drop-down arrow and choose a type of repository from the menu (Figure 12.14). The name is listed on the preferences (hidden below the menu in the figure).

3. Click OK to close the Preferences dialog box and save the server setting.

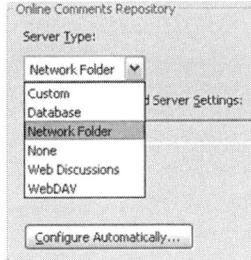

Figure 12.14 Specify an online repository for convenience

Participating in a Review in Adobe Reader 7

An enabled file distributed for review from Acrobat 7 Professional displays certain features in Adobe Reader 7, as seen in Figure 12.15. These features are not available at all times when a file is opened in Adobe Reader, as the commenting is enabled on a file-by-file basis.

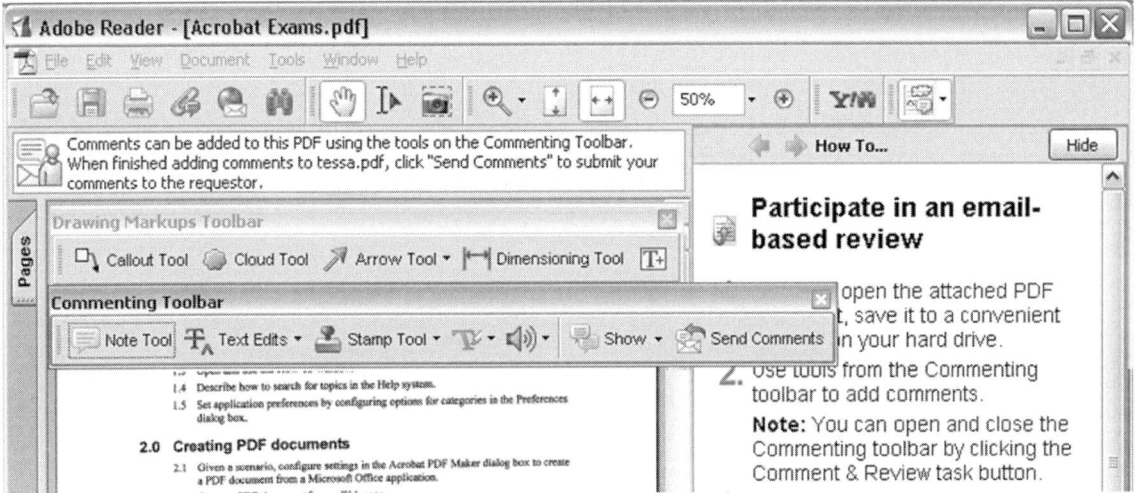

Figure 12.15 Additional features in an enabled PDF file

The interface includes:

- The Document Message Bar, explaining that the commenting features are enabled.
- The Commenting toolbar, as well as the Drawing Markups toolbar (if specified when initiating the review) overlay the program window.
- A Comment & Markup Task button shown on the Adobe Reader toolbar.
- The Comments tab is included in the Navigation tabs.
- The How To pane, listing instructions for participating in an email-based review.
- The Tools menu contains commands for opening and closing the enabled toolbars.

Returning and Receiving Comments

When you have completed adding comments to a file in a document review, click Send Comments ☒ on the Commenting toolbar to open the Send Comments:[name of file] dialog box.

Make any necessary changes to the email address, subject line, or content of the email, and click Send. The dialog box closes and the file is sent. As with initiating a review, depending on your system's configuration, you may see additional dialog boxes – follow the prompts to send the file.

Receiving Comments

If you are managing a review the comments are returned as a PDF or FDF file; you can either open the file directly from the email message, or save it to your hard drive.

Follow these steps to open a commented file and incorporate it into a master document:

1. In your email program, double-click the PDF attachment to the received email. The file uses the name of the original file, unless the person commenting saved the file with a different name before returning comments to you.

2. Acrobat 7 opens automatically, and shows a Merge Comments? dialog box, asking if you want to merge the comments into the original document used for the review or to open a copy only (Figure 12.16).

3. Click Yes to close the dialog box and load the file.

4. The original review document opens showing the Comments pane. The comments are placed on the document and listed in the Comments pane (Figure 12.17).

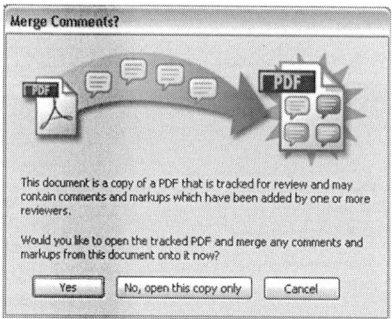

Figure 12.16 Acrobat tracks a document's status

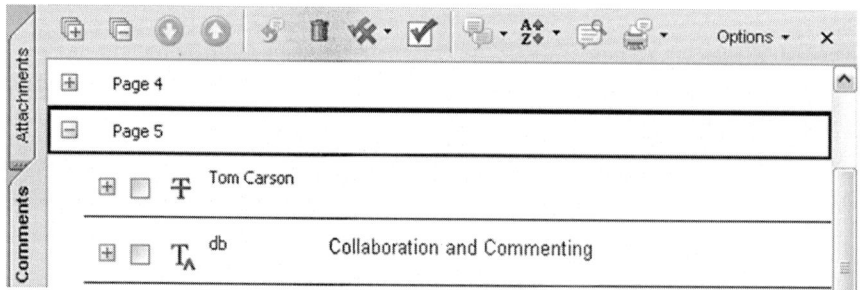

Figure 12.17 The Comments pane is shown automatically

Grouping Comments

It is common to see an object circled on a document and an arrow pointing to the same object. Amalgamate comments into a single grouped comment to make them easier to work with. On the page, Ctrl-click to select the objects, and right-click to open the shortcut menu and choose Group. The objects are combined into one unit, and display a distinctive grouped icon ⬚ on the Comments pane.

Manipulating the Comments List

Comments can be manipulated in several ways, useful when working with a large number of comments. You can sort comments or apply a filter to reorder and display part of the list. If you are looking for a specific term, Acrobat lets you search comments as well.

Sorting Comments

Imported comments are sorted using the default sort by page method. You can use the same navigation tools, such as collapsing or expanding all headings in the Comments pane, regardless of the sort method used.

Sort comments using a variety of heading types depending on your workflow. Sorting shows all the comments in your document.

To choose a different sorting method:

1. Click the Sort By button on the Comments pane to open its drop-down menu; the current sorting method is checked on the menu.

2. Click a sorting method from the list.

3. You see the content is rearranged on the Comments list (Figure 12.18).

Figure 12.18 Sort comments according to your workflow

Filtering Comments

Applying a filter in the Comments list limits the display of comments to only the selected options. Follow these steps to apply a filter:

1. Click the Show button to open its menu.

2. Choose a filter option and then a filtering method (Figure 12.19).

3. Click OK to close the information dialog box that displays.

4. On the Comments pane, the filter is applied to the comments list; you also see a status message regarding the filter.

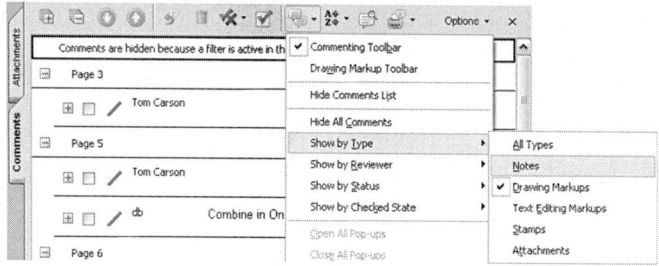

Figure 12.19 Choose filtering options

Searching Comments

If you are working with a long document or one containing a large number of comments, you can search for particular content in a comment. Follow these steps to search comments:

1. Click Search comments 💬 on the Comment pane to open the Search PDF window at the right side of the program window.

2. Type the word or phrase to use as the search term, as well as other search options in the Search PDF window.

3. Click Search Comments.

4. Acrobat processes the file and displays the results.

5. Click the result in the Search PDF Results list to select the corresponding comment in the Comments pane (Figure 12.20).

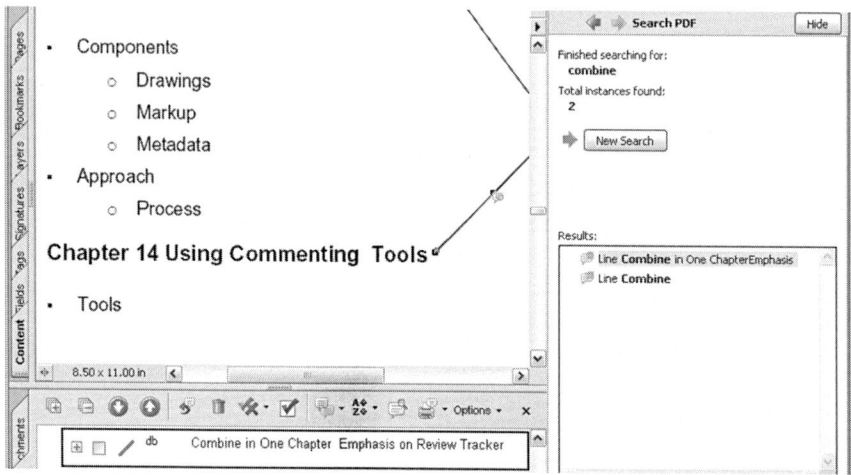

Figure 12.20 Search comments for text strings

Organizing Comments

In addition to offering a number of ways to order and present comments visually in your document, you can also use a number of tools to organize the material to meet varying demands in your workflow. For your own use, you can use a check mark system to differentiate one group of comments from another, add a reply to any existing comment, or assign a review status that is shown with the comments in a review.

Using Check Marks

Acrobat offers check marks that you use for organizing your work internally. That is, the check marks are seen on your copy of the document on your computer, but are not included in the file itself if transmitted to others. You can use check marks in a number of ways, such as identifying tasks completed or outstanding, comments that have been processed, or read, search results, and so on.

Click the check box to the left of a comment's listing in the Comments pane to add a check mark, or select one or more comments on the Comments pane and click the Checkmark icon ☑ on the Comments pane's tools. Remove a check mark by deselecting the check box, or selecting the comment/comments and clicking the check mark icon on the Comments pane's tools.

When you have used check marks, you can also use them as a method for sorting or filtering comments. Figure 12.21 shows the Comments list with checked and unchecked filters applied.

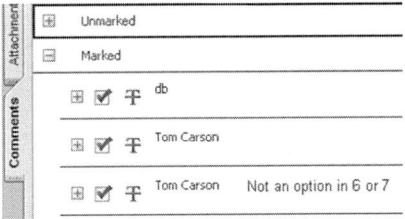

Figure 12.21 Use check marks as a method of sorting or filtering

Replying to Comments

Using comment replies is a convenient way of responding to others' comments without unnecessarily cluttering up a document, or using an unmanageable number of comments. Comments containing replies are seen if you perform multiple rounds of reviewing, as well as displaying in summary reports.

To reply to a comment, follow these steps:

1. Select the comment on the Comments list.

2. Click Reply 🖰 on the Comments pane's tools.

3. A blank row is added to the active comment, showing a text field over a yellow background. Type the reply in the text field (Figure 12.22).

4. Continue adding other comment replies depending on your workflow. If you collapse a comment containing a reply, a notation is shown in the Comments pane.

On the document page, you see the reply overlays the original comment (Figure 12.23).

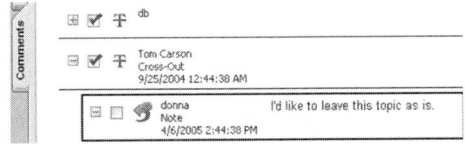

Figure 12.22 A reply is identified on the comment's row

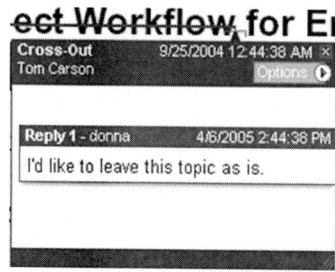

Figure 12.23 A reply is attached to the original comment

Setting a Review Status

Check marks are used internally; if you want to communicate with your review group, use a Review status or state for the comment instead. Review states can be used to represent different things, depending on your project's needs.

Follow these steps to assign a status:

1. Select the first comment for assignment.

2. Choose Set Status > Review > and an option from the Comments pane's tools.

3. The Review state is shown below the comment details, and includes the selected state and the time/date.

On the document, a comment with a Review status displays the status in the tooltip (Figure 12.24).

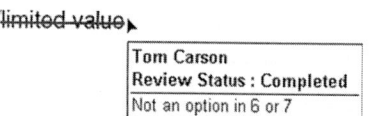

Figure 12.24 Read the Review status in the tooltip

Modifying a Reply

Acrobat defines a reply as a modified Note comment. Right-click the reply to open the shortcut menu and choose Properties.

-   In the Note Properties dialog box, you can change most of the note features, such as the name or subject on the General tab.
-   On the Appearance tab, however, you cannot choose an icon, as the reply is integrated into the original comment and does not show its own icon on the document page.

Migrating Comments

In certain circumstances, comments may not import properly into a master document. This can occur when you have performed a review, integrated comments, and then revised the document and recirculated it again.

When comments are migrated, instead of looking for a matching page location Acrobat looks for matching word groupings and structural elements. Text comments referencing words are placed where Acrobat finds a matching word grouping; visual comments, such as drawing markups, are placed at the matching structural location on a page. Acrobat places a comment on the same page in a revision as in the original. However, if the original page has been removed, the comment is placed on the last page of the document.

Like regular integrated comments, migrated comments can also use a Status. Select the comment, and then choose Set Status > Migration > and choose an option from the menu. A comment with an applied Migration status shows a label on the Comments pane (Figure 12.25).

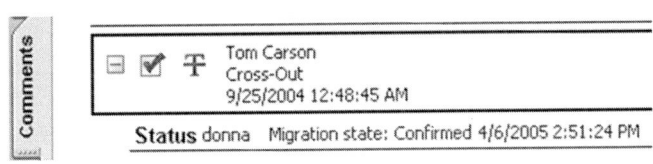

Figure 12.25 Use a Migration status in a revised document

Summarizing and Printing Comments

Once a review is complete and the issues addressed, the final stage depends on the workflow and need for maintaining records. In a casual review, you are most likely to finish the project, and then delete the review from the Tracker. In more formal reviews, or those that maintain records for legal, government, or other purposes, Acrobat offers ways to maintain and store a summary document.

To create a summary document, follow these steps:

1. Choose Summarize Comments from the Options menu in the Comments pane to open the Summarize Options dialog box.

2. Choose the option for the document, including the page layout, paper and font size, and a sorting option.

3. Click OK to close the dialog box and generate the report, which opens in Acrobat; save the PDF file.

Automatic Options

Rather than opening dialog boxes and choosing preferences, you can select one of two ways to generate files automatically. On the Comments pane, click Print Comments, and choose Print Comments Summary to send the document and comments directly to your printer using Acrobat's default program settings. If you want to save a copy of your comment report, choose Create PDF of Comments Summary to build a summary document using the program defaults.

Other Comment Export Options

Comments can be exported in other formats or as subsets:
- Comments can be exported to other programs. Click the Options menu and choose Export Comments, and choose an option. Comments can be exported to Word, AutoCAD, or saved as an FDF or XFDF file. Read about returning comments to AutoCAD in Chapter 6.
- Export a selected group of comments from a file. Select the comments, then click the Options menu and choose Export Selected Comments. A Save As dialog opens; name and save the file, stored in FDF format.

Types of Summary

Choose a type of comment summary that is best for the characteristics of your document and how you intend to use the summary report. The comment summary options are shown in Table 12.2.

Table 12.2 Comment summary types

Summary Type	Icon	Common Uses
Document and comments with connector lines on separate pages		Use to work onscreen; set the two pages side-by-side in Acrobat and clearly see the content
Document and comments with connector lines on single pages		Use if both reading online and printing
Comments only		Use for text-heavy documents; generates a printed list of comments to compare against revisions
Document and comments with sequence numbers on separate pages		Use for text-heavy documents when printing the reports

Integrating Comments into a Source File

The Text Edit comments can be exported as actual edits to the source Word document if you are working in Windows and the source document was created as a tagged PDF file in Windows 2000 or newer using PDFMaker.

Follow these steps to select and export comments from Acrobat:

1. Click the Options button on the Comments pane and choose Export Comments > To Word from the menu.

2. Microsoft Word opens, and Acrobat displays a large dialog describing the integration process. Read through the dialog and click OK.

3. In the Import Comments from Adobe Acrobat dialog box that opens next, choose the files and comment import options (Figure 12.26). The name of the open PDF file is listed in the Take comments from this PDF file field. Click Browse to locate and select a different file.

4. Click Browse to locate and select a file into which the comments will be imported. The name of the file displays in the Place comments in this Word file field.

5. Select comments to import. You can choose among all comments, those with check marks, Text Edit comments only, or specify a custom filter, such as filtering by author.

6. To display the edits in the Word document, leave the Turn Track Changes On Before Importing Comments check box selected.

7. Click Continue.

Note: If you want to use a manual selection method for choosing comments for export, use check marks in the Comments pane before starting the comment integration process. Otherwise, Acrobat offers other methods as you work through the dialog boxes.

Acrobat processes the comments, and the Successful Import dialog lists the numbers and types of comments imported (Figure 12.27). You can check each Text Edit comment before it is committed to the file, or click Integrate Text Edits on the Successful Import dialog.

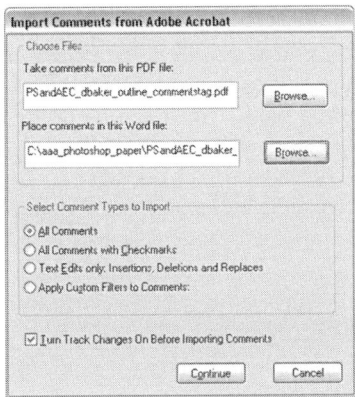

Figure 12.26 Select comment import options

Tip: If you are using a renamed file, an information dialog box warns that comments may not be placed correctly. Click Try Anyway.

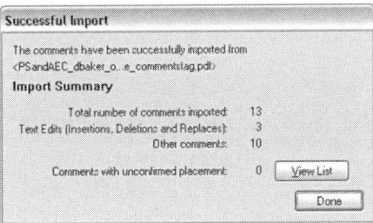

Figure 12.27 Import status is displayed

Flattening Comments

ARTS PDF Tools (www.planetpdf.com) has a tool called a *flattener*. Once you have filled out the form, Flattener will flatten the form fields to the base layer of the document, which allows you to submit the document for review without adding security. When it comes back with a big approved stamp, you can flatten the stamp to the base layer again and add this to the Electronic Owner's Manual.

In the ensuing Adobe Acrobat Comments dialog box, you can choose to accept each edit individually, or accept them all and import them into the source file. The edits are made automatically based on the content and type of the original Text Edit comments added in Acrobat (Figure 12.28).

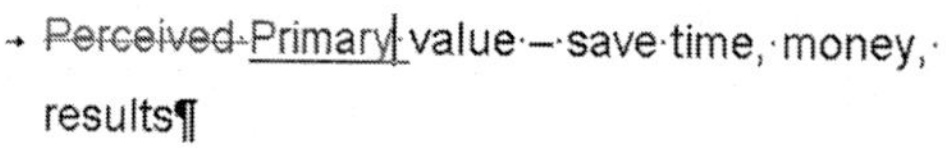

→ ~~Perceived~~ Primary value — save time, money, results¶

Figure 12.28 Corrections are automatically integrated

Summary

In this chapter we examined the extensive functionality of Acrobat's commenting and markup processes, as well as how to organize and manage a review cycle.

Acrobat offers numerous commenting tools, which can be customized to meet your workflow needs; you can even create custom tools as Stamp comments. In addition to comments, Acrobat also offers specialized viewing tools, often used when reviewing a document.

Initiate reviews using Acrobat's wizards, and track and manage your reviews through the Tracker. Manage, filter, sort, and create reports from the Comments pane.

Project

A submittal is a perfect example for review and commenting. Instead of six paper copies, we are going to send one electronic copy to the architect or engineer for approval.

Task 1: Preparing the submittal

1. Open the file in the **ch12_project** folder named **Submittal Package.pdf**, which includes a fillable PDF form.
2. Complete the form based on the drawing and linked manual. We have included a CAD drawing and an owner's manual that is approximately 50 years old.

Task 2: Adding password security

Check the configuration of the signature field placed to the right of the SIGNED: label on the bottom of the Letter of Transmittal form (page 1 of the project file) and then add the password:

1. Select the Digital Signature tool on the Forms toolbar, a subtoolbar of the Advanced Editing toolbar.
2. Double-click the Signature1 field at the bottom of page 1 of the project file to open the Digital Signature Properties dialog box.
3. Click the Signed Tab. Note that the Mark as read-only radio button is selected, which means that, after the file is signed, all fields on the form become read-only.
4. Click OK to close the dialog box.
5. Choose File > Document Properties > Security to display the Document Security settings.
6. Click the Security Method drop-down arrow and choose Password Security.
7. In the Password Security dialog box, click the Use a password checkbox in the Permissions area of the dialog box; the remainder of the dialog box's fields become active.
8. Click the Printing Allowed drop-down arrow and choose High Resolution; click the Changes Allowed drop-down arrow and choose Commenting, filling in form fields and signing existing signature fields.

9. Enter a password and click OK; follow the prompts to confirm the password and close the dialog box.

10. Save the file.

11. Sign the document. You see once it is signed, the form fields are read-only, which is a practical method of ensuring the file's contents are secure. The only issue with this method is that the document cannot be added to another document, such as when building an Electronic Owner's Manual.

Task 3: Initiating the review

Send the file for commenting to a reviewer using this method:

1. Choose File > Send for Review > Send by Email for Review (or another method of selecting the command) to open the wizard dialog box.

2. In the Send by Email for Review wizard, follow the prompts to specify the review document in the first pane of the wizard; click Next.

3. On the second pane of the wizard, click Address Book and select the recipients' email addresses, or type the addresses in the field.

4. Also on the second pane of the wizard, click Customize Review Options to open the Review Options dialog and check the Display Drawing Markup Tools for this review check box. You will see in the dialog box that Review with Adobe 7 Reader is chosen by default. Click OK to close the dialog box, returning to the wizard.

5. Click Next to display the third pane of the wizard. Preview the default subject and text intended for the invitation. Modify the content if you like.

6. Click Send Invitation to send the email. If necessary, follow your email client's prompts or dialogs.

It is simpler to exchange files among two or more people. If you have two email addresses, or two computers, you can try sending the invitation to yourself. However, since Acrobat monitors the master file which is already on your computer, you may not be able to open an invitation copy.

Task 4: Working with commenting tools

Have fun with the commenting tools – in a constructive way, of course.

1. Experiment with and apply a custom stamp created from an image file; a recorded sound comment, an assortment of other comments, and some drawing markups.

2. Display page 2 of the project file, showing the drawing. Choose Tools > Measure and measure the length of the Scale; evaluate to see if the drawing and embedded scales match (the tool is available only in Acrobat 7 Professional.)

3. When your file is "suitably" marked up, click the Send Comments button on the Commenting toolbar or the Document Message Bar to return the file to the review initiator.

Task 5: Review tracking

The files returned by email can be incorporated into the master document immediately, or saved to your hard drive and added at any time.

To open and use the file from the email:

1. Double-click the file attachment to open the project's master file and accept the comments for merging into the master file.
2. Click the Comments tab to display the Comments pane.
3. Experiment with the functions in the Comments pane, such as the sort, filter, and set status commands.
4. Choose Options > Tracker from the Comments pane to open the Tracker window.
5. Click Manage and investigate the optional commands
6. When you have finished experimenting, select the project file in the My Reviews listing in the Tracker and click Remove. Confirm deletion.

13

Using Other Features and Functions

Acrobat includes a range of advanced features that are a bonus for AEC. Although many practicalities and applications of these features may not be readily evident, we will show you their importance. The features range from document contents to ways of streamlining your workflow to using multimedia.

In this Chapter

In this chapter you see how to apply several of Acrobat's advanced features to your workflow, including:

- Adding and manipulating metadata and other document properties
- Using video, multimedia, and Flash
- Creating presentations
- Building batch scripts
- Writing and using Droplets.

Managing Document Data

Acrobat allows extensive data to be added to a PDF file. On first thought, you may have a difficult time understanding how additional data are of value to you in AEC. Here is an example: imagine having 10,000 important images for a project, and then imagine being able to search for a particular image based on a part number, a date, or a combination of variables.

Metadata is descriptive information about a file that is searchable. Acrobat uses Adobe's eXtensible Metadata Platform (XMP) to embed metadata into a file. Applications that support XMP share the information across file formats, databases, and platforms. For example, metadata added to a source image in Photoshop is converted along with the image data to PDF. It takes some time to develop a structure for defining and including metadata, but it will take only seconds to find a specific image in an emergency.

Modifying Document Properties

Additional metadata categories and specific values can be added either in the source program, such as Photoshop CS, or from Acrobat in the Document Properties dialog box.Follow these steps to modify the properties:

1. Choose File > Document Properties to open the Document Properties dialog box.

2. Add additional information you want to use for searching and organizing in the fields, such as Subjects or Keywords (Figure 13.1).

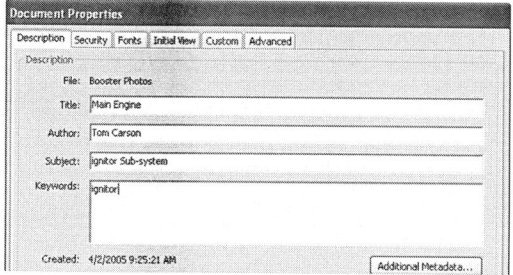

Figure 13.1 Add metadata for cataloging and searching

3. For images, you can add more data using the XMP format. Click the Additional Metadata button to open the XMP dialog box, add the information, and click OK to return to the Document Properties dialog box.

4. To add more properties, click the Custom tab. Here, you can add properties and values to customize your data storage further (Figure 13.2).

5. Click OK to close the Document Properties dialog and save the file.

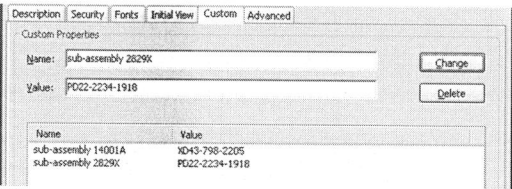

Figure 13.2 Custom values enhance searchability

Viewing Metadata

Visio and AutoCAD drawings can include embedded metadata, which is derived from custom information added in the source program. To view custom metadata, choose Tools > Object Data > Object Data Tool. Click to select an object on the page containing metadata, indicated by a crosshairs cursor, to open the Object Data dialog box.

Here, you will see the embedded data and its values (Figure 13.3).

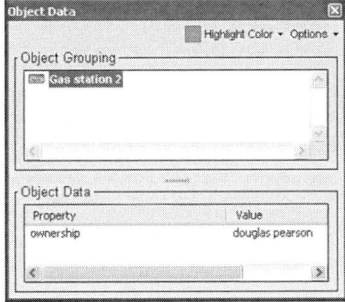

Figure 13.3 Some file formats include embedded data

The Options menu on the dialog box includes commands such as zooming to a selected object, counting similar objects, or copying the object's content to the clipboard.

Using Media in a PDF Document

Suppose the mechanically brilliant old codger that kept the plant running is going to retire shortly. The plant has some machinery that only he can fix. If you are smart, you will use your digital camera with sound to capture him explaining how to fix the machinery. Then you can embed the movies, images, and sound into a PDF file created in either Acrobat 6 or 7. Remember that Acrobat 5 would only link to a movie, but it couldn't be embedded. Once the PDF file is finished, store it in the owner's manual for that piece of equipment for posterity.

Movies can be embedded or linked to your document. They can be set to play on page opening by specifying the trigger attached to the page. You can click a marquee to activate the movie, or initiate it by clicking a link, bookmark or button. Sound can be added by linking to a sound file or adding a sound file as an attachment.

Multimedia Preferences

Set multimedia preferences and trusts in the program preferences. Choose Edit > Preferences > Multimedia. In the Player Options, click the drop-down arrow and select a player option – the choices do not include the particular player's version (Figure 13.4).

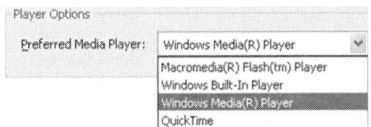

Figure 13.4 Choose player preferences

To set permissions, choose Edit > Preferences > Trust Manager to display the settings. Click the Change Permission for selected multimedia player to drop-down arrow and choose an option (Figure 13.5). A trusted document means it is included in your list of trusted documents and authors.

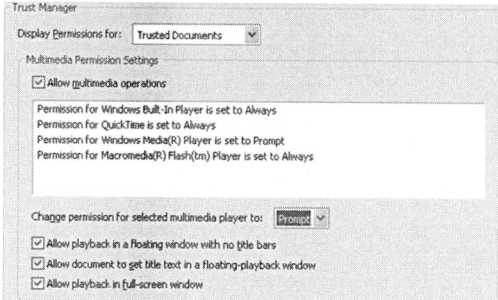

Figure 13.5 Set Trust Manager options for players and formats

You can select different options for the listed players. Select a player and set its permission level:

- **Always** plays content in the player at all times
- **Never** prevents the player from being used
- **Prompt** asks for a decision when a nontrusted document is open that contains media, and you are prompted to add the document to your list of trusted documents or authors.

Adding a Multimedia File to a PDF Document

In Acrobat 7 Professional you can either embed or link a movie. An embedded movie is integrated into the PDF document; a linked movie has a link pointing to the original location of the movie. The movie must be embedded to use controls.

Follow these steps to add a movie:

1. Select the Movie Tool ▦ on the Advanced Editing toolbar.

2. Double-click the page where you want to place the upper left of the movie or drag a marquee and release the mouse to open the Add Movie dialog box (Figure 13.6).

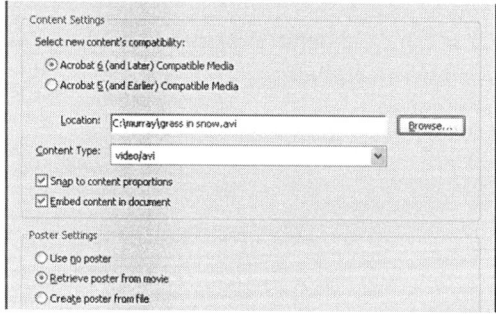

Figure 13.6 Choose settings for placing the movie

3. Click Acrobat 6 (and Later) Compatible Media and click Browse to locate and select the movie. When you select a file, a content type is assigned that determines the player needed for viewing automatically.

4. Select additional options to embed content in document (which includes the movie in the PDF file); Snap to content proportions maintains the frame size when the movie plays.

5. Choose a poster option. A poster is a placeholder image that displays on the page when the movie is not playing. You can choose no poster; retrieve poster from movie, which uses the first frame as a static image; or create a poster from a file. If you choose the last option, click Browse to locate and select the image file to use for the poster.

6. Click OK to close the dialog box.

7. To test the movie, click the Hand tool, and then click the movie's frame to start playback.

File Size and Formats

In order to embed media clips, use renditions, use a different file as a poster, or use a range of content; you can only use the Acrobat 6-compatible option in the Add Movie dialog box.

Be aware of file sizes when embedding movies into a PDF document, as a movie can dramatically increase a file size, depending on frame rate and frame size. If you are using a movie in-house or internally, consider linking to it; for movies exported or emailed, embed the movie file to prevent corrupting the link between the media file and the PDF.

You can use many different media formats in a PDF document, ranging from Flash, to common video formats such as AVI and QuickTime, Windows Media Player formats, and Windows player formats.

Adding Sound Files

Adding a sound file is a similar process. Choose the Sound Tool 🔊 on the Advanced Editing Toolbar. Click the page where you want to attach the file, which is usually invisible, unless you want to use a poster to identify a sound button. The Add Sound dialog box opens and displays the same options as those for adding a movie, except for the Snap to content proportions, which does not apply to a sound file.

Using Movie Renditions

You can use a number of versions, or *renditions* of a movie. Use them for distribution when you are unsure of the player versions or types your viewers are using, or if you want to have both high- and low-quality versions of a movie.

Double-click the movie on the page with the Movie tool to open the Multimedia Properties dialog box where you see movie settings and options (Figure 13.7).

Choose from these settings:

- The Multimedia Properties dialog box opens with the Settings tab, which lists the Annotation Title, the name Acrobat assigns to identify the object.
- The List Renditions for Event field shows the Mouse Up action as the default movie playback action; other actions are available from the drop-down menu.
- The initial movie added is shown as the first rendition. To add more renditions, click Add Rendition, locate and select the versions you want to use.

You can change the appearance of the movie on the page, such as its border, by making selections from the Appearance tab (Figure 13.8). Customizations include stroke width, style, and the type of border, as well as color options.

Click Edit Rendition to open the Rendition Settings dialog box. In this dialog box you can set a range of custom options, such as using floating windows or movie playback controls (Figure 13.9).

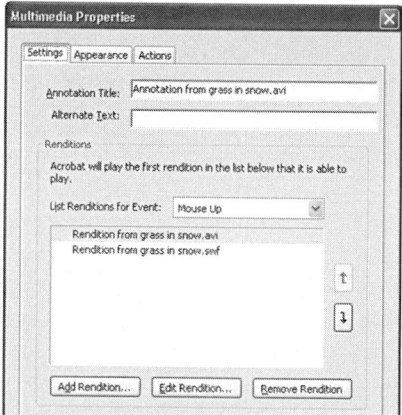

Figure 13.7 Choose renditions in the dialog box

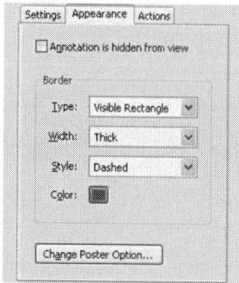

Figure 13.8 Select custom options for renditions

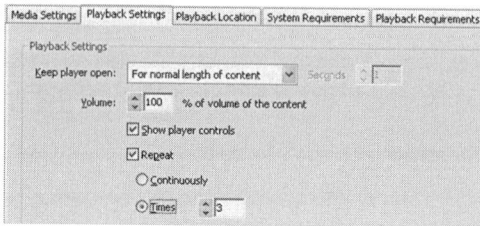

Figure 13.9 Define the appearance of the movie on the page

Customize your movie using the five-tabbed Rendition Settings dialog box:

- Media Settings. Make a rendition accessible to JavaScript; modify basic movie options, like those found in the Add Movie dialog box.
- Playback Settings. Define looping for the movie, specify players, and controls (**Figure 13.10**).
- Playback Location. Define if the movie plays in a floating window or full-screen instead of a defined area.
- Systems Requirements. Specify languages and playback requirements, including subtitles and screen resolution.
- Playback Requirements. Read a summary of the settings chosen in other tabs.

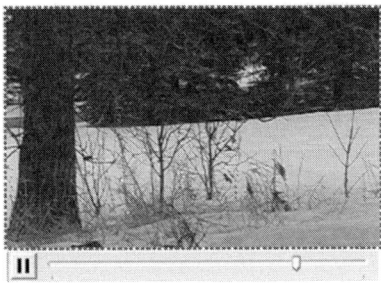

Figure 13.10 Customize a movie playback using controls and other features

---

Multimedia Actions

There are three multimedia-specific actions in Acrobat:
- Play a Sound. This action embeds the sound in the PDF document in a cross-platform format that plays in Microsoft Windows and Mac OS.
- Play Media (Acrobat 5 Compatible). This action plays a QuickTime or AVI movie created as Acrobat 5 compatible. A media object using Acrobat 5 Compatible options is automatically embedded in the PDF document.
- Play Media (Acrobat 6 and Later Compatible). This action plays a movie created as Acrobat 6 or 7; the media object must already be embedded in the document to use the action.

Presentations

PowerPoint changed presentations for better or worse, depending on the quality of the presentation and the presenter. Acrobat is changing presentations again, especially for engineers. PowerPoint includes a range of design backgrounds, transitions and effects, which are converted along with the file when converted with the PDFMaker.

Assembling a Presentation

You can use converted PowerPoint presentations, other converted PDF documents, or assemble a binder document from multiple sources. Acrobat lets you add effects, configure transitions, and specify the trigger for advancing pages.

Follow these steps to build a presentation:

1. Choose Document > Set Page Transitions to open the Set Transitions dialog (Figure 13.11).

2. In the Set Transitions dialog box, choose a transition effect, and a speed.

3. Specify the navigation method. Auto Flip automatically advances the slideshow based on the number of seconds you specify. If you do not select Auto Flip, the user moves by mouse click or keyboard command.

4. Choose the page range or leave the default, which is the entire document.

5. Click OK to close the Set Transitions dialog.

6. Test the presentation by clicking the Full Screen View 🖳 button on the status bar at the lower left of the program window.

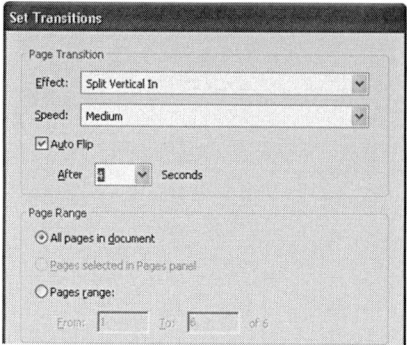

Figure 13.11 Specify the presentation's features

The transitions are visible only in Full Screen view. To specify the document opens in this view, choose File > Document Properties > Initial View. Click Open in Full Screen mode in the Window Options section.

Presentations Using Picture Tasks

Acrobat displays the Picture Tasks plug-in if you open image PDF files that originated as JPEG images. Use the tasks to create visual elements quickly such as multiple prints and presentations. Picture Tasks are a subset of Adobe's Photoshop Album program.

To create a Picture Tasks slideshow, follow these steps:

1. Click the Picture Tasks button and choose Export to Slideshow to open the Export to Slideshow dialog box.

2. Select the images for exporting to the slideshow from the thumbnails view at the left of the dialog box (Figure 13.12).

3. Specify a title for the slideshow, the slide duration, and the transition to use. You can also specify a background sound score for the presentation.

4. Click Export to create the slideshow; save the finished presentation PDF file.

How Big is Big?

Tom created the 'World's Largest Computer Presentation'. At over 51,000 pages, the presentation includes a few PowerPoint slides, and the rest are PDF pages generated from AutoCAD, MicroStation, Word, and Excel, scans, and converted microfilm. The presentation includes a few thousand CAD drawings converted to PDF format, and more than 1500 geo-referenced PDF maps with embedded data. Using a system of links, buttons, and bookmarks, this enormous collection of files navigates like a single file.

The presentation includes the *Marshall Star* newsletter, which is NASA's oldest continuous publication. The publication at NASA Marshall Space Flight Center covers 44 years in more than 2000 issues and over 16,000 pages.

The search capabilities blow people away. Often around Huntsville, Alabama, someone will test Tom by asking if their father or uncle was in the Marshall Star. They are often surprised that their relative was a champion bowler on the Marshall team, or that they won an award for a new invention.

The World's Largest Computer Presentation ends with an animated, interactive 3 – D page created by Bentley in MicroStation and Acrobat. The crowd always says, "Wow! I never knew PDF could do that!"

Note: You can specify program preferences for presentations. Choose Edit > Preferences > Full Screen and choose a transition. Select Ignore All Transitions to hide transitions in any document you open.

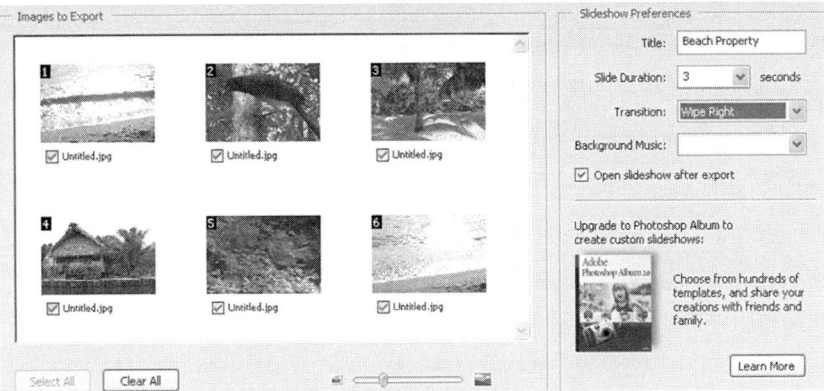

Figure 13.12 Use Picture Tasks for a quick slideshow

Using Picture Tasks is a simple way to build a slideshow. In addition to adding the presentation features, the process also creates and applies a control panel to the finished slideshow (Figure 13.13). If you build a slideshow manually, be sure to include links, bookmarks, or buttons to control the navigation.

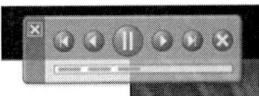

Figure 13.13 Picture Tasks creates a control panel

Batch Processing

Many things in life are simpler if you can work in batches, and working in Acrobat is certainly no exception. You can use any of the existing scripts as is or customize them, or write your own scripts.

Decide what you need to add in the script before starting a big project to prevent manually repeating simple tasks that could be incorporated into the batch sequence. Before using a batch script, put the files you want to batch in a separate folder to keep track of them. Test the batch sequence on a sample file to make sure you are pleased with the outcome.

We will show you how to construct a simple batch script for adding document information to a series of image PDF files. Follow these steps to construct and run the batch script:

1. Choose Advanced > Batch Processing to open the Batch Sequences dialog box and click New Sequence. Type a name for the new sequence in the small dialog box and click OK.

2. The Edit Batch Sequence [name] dialog box opens. Click Select Commands to open the Edit Sequence dialog box (Figure 13.14).

3. Click an arrow to open the category of action and a specific action and then click Add to move the action to the list at the right of the dialog box.

4. Click the action in the right column to open a dialog box used to specify the features of the command. Of course, the available settings depend on the command selected.

5. Continue to add commands as required, then click OK to close the dialog box and return to the Edit Batch Sequence [name] dialog box.

6. Specify the options for running the command, such as the files the sequence is applied to, and when the commands are applied (Figure 13.15). Click Output Options to configure the processed files further, such as saving them in another image format.

7. If you are ready to use your custom sequence or an existing batch sequence, click Run Sequence.

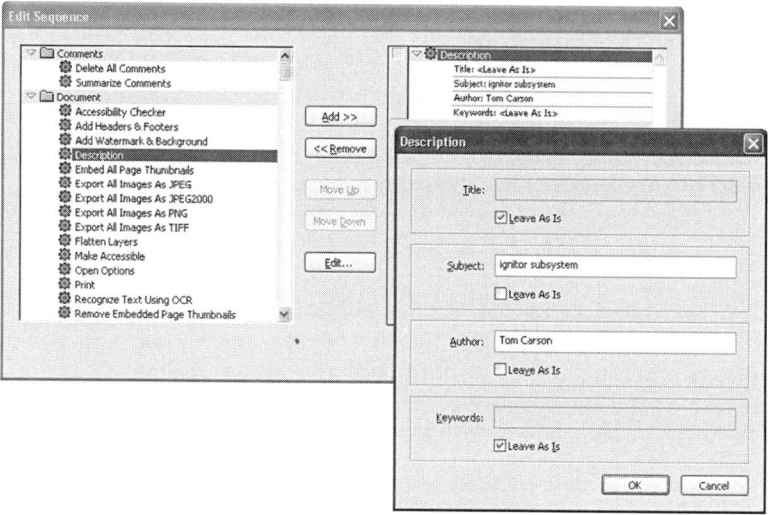

Figure 13.14 Choose commands and specifications for the batch script

Security Alert

Set Security to No Changes is probably the most widely used batch process, but can be a problem. Choose Edit > Preferences > Batch Processing, and notice that the Default Security Method is Do not ask for Password.

What happens when you set Password Security without a Password? Just that: security is set without a Password. Although Acrobat 5 and 6 allowed this action, Acrobat 7 opens the dialog for password security and asks you to add the password.

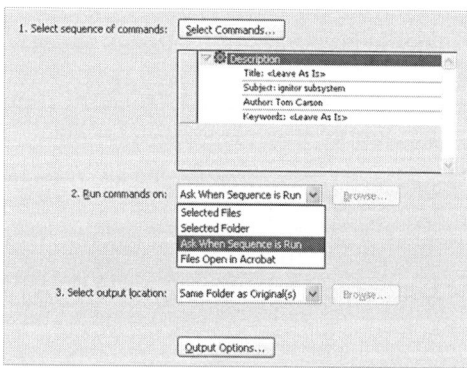

Figure 13.15 Set options for running scripts and storing output

Note: Assign an Interactive Mode by clicking the solid gray box to the left of the command's name. When the batch runs, as each file reaches the specified command the corresponding dialog box opens for you to specify settings. Different files might have different keywords, for example, which could be added on a file-by-file basis.

Droplets

A Preflight Droplet is an application you build in Acrobat for testing and evaluating files. Although a droplet is characteristically used to evaluate documents for commercial printing, there are many options that can be applied to an AEC project. For instance, you may have hundreds of images associated with a project that are combined into a binder files. You can construct and run a droplet that will list the images by name in a text file.

Follow these steps to construct the Droplet:

1. Choose Tools > Print Production > Preflight to open the Preflight dialog box.

2. Click Options on the Preflight dialog and choose Create Preflight Droplet. The Preflight: Droplet Setup dialog box opens (Figure 13.16).

3. Choose the profile from the Run Preflight check using: drop-down menu.

4. Click Settings to open the Preflight:Report Settings dialog box, also shown in Figure 13.16 and specify the type of report and level of detail. Click OK to close the dialog box.

5. Click Save; the Save Droplet As dialog box opens. Specify a name and storage location and click Save to create the Droplet. The simplest location to store a Droplet is on your desktop for ease of access.

When you want to test a file, locate and select the file on your computer. Then drag it to the Droplet icon if it is stored on the desktop (Figure 13.17). Acrobat starts, tests the file, and generates the report, listing the images and their characteristics.

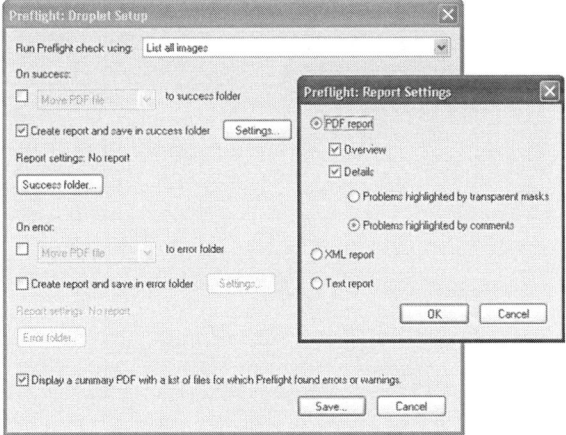

Figure 13.16 Configure the content for the Droplet

List all
images.exe

Figure 13.17 Leave the icon on the desktop for easy access

Once created, you cannot modify a Droplet, as it is an executable file. You have to rebuild the Droplet to revise its function.

> **Tip:** You can choose separate folders for success and error results if you like – separate folders are more commonly used in preflighting documents for press.

Optimizing PDF Files

The number of ways you can add to a file's size are as numerous as the commands available in Acrobat, ranging from adding and removing movie file iterations to using images with resolutions higher than necessary.

Acrobat 7 Professional includes the PDF Optimizer to evaluate and streamline your files. The default settings are for Acrobat 5 compatibility; choosing another version produces a Custom preset for the command.

Follow these steps to use the PDF Optimizer:

1. Choose Advanced > PDF Optimizer to open the dialog box, and click Audit space usage at the upper right to view the document's report listing sizes of the file's elements in percentages of the entire document size and in bytes (Figure 13.18). Click OK to close the audit report.

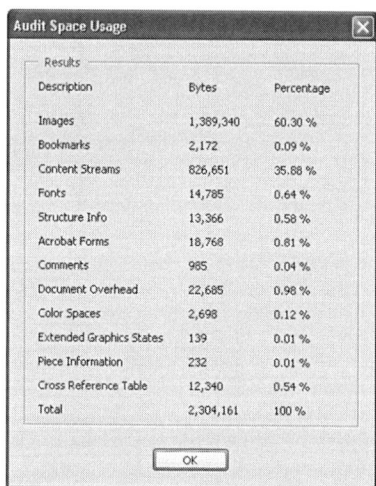

Figure 13.18 Acrobat shows you the file usage breakdown

> **Note:** Optimizing a signed document will invalidate the signature. (Read about signatures in Chapter 10.)

2. Click a label in the left column on the dialog to display commands in a set of panes (Figure 13.19). Read through the lists and select/deselect items to apply in your document:

- **Images.** Compression types, quality, and downsampling values.
- **Scanned Pages.** Click the compression and quality check box and use filters to clean up scanned pages, such as despeckle or halo removal. Choosing Adaptive Compression options disables the settings on the Images pane.
- **Fonts.** Check the listed fonts and unembed system and common fonts.
- **Transparency.** Select settings such as resolutions for text, line art, gradients, and transparency flattening.
- **Discard Objects.** Select objects that can be deleted, such as form content, layers, or comments.
- **Clean Up.** Choose other clean up details, such as removing unused links or bookmarks.

3. Click OK to close the dialog box (you can click Save to save the settings for future use). The Save Optimized As dialog box opens; click Save to overwrite the original file or save it with another name.

Printing

PDF is a major standard for the printing industry. Acrobat 7 Professional is used extensively by the print professional for a wide range of prepress functions. It is rare that an engineer needs to know these features, but they are there if you ever have to send a press-ready advertisement to a major magazine.

Tom's advice on high-end printing: find yourself a Donna! Donna's advice on high-end printing: experiment and learn.

Resaving a File

One of the simplest ways to reduce file size is to save a file as itself. Whenever you save a document, it saves an iteration. After time, the file can become very large. Choose File > Save As, leave the file's name as is, and click Save. Click Yes to the overwriting prompt to consolidate and resave the file.

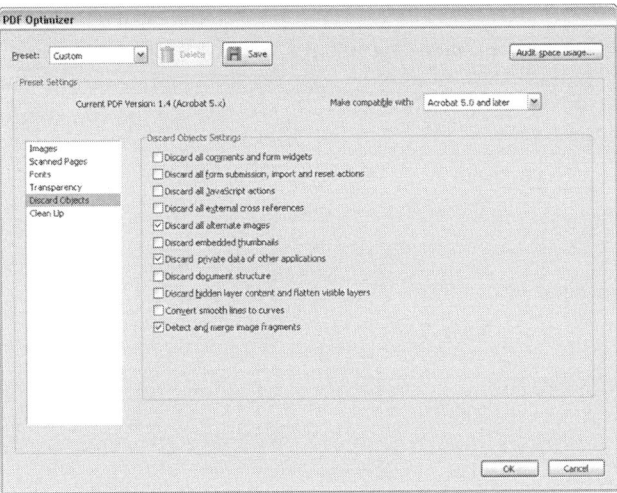

Figure 13.19 Save file size using the PDF Optimizer

Summary

In this chapter you learned about document properties and metadata. Remember that metadata and using properties such as keywords are important for efficient searching in a large collection. Acrobat PDF files can use embedded or linked media, and you saw how to add and configure video, Flash, or sound files.

You saw two ways to create presentations, either manually or using a plug-in. The rest of this chapter was all about ways to save time, increase your efficiency, and use automated features, such as batch files, automated sequences for performing a range of commands. You learned about Droplets, which are executable files that trigger an evaluation process in Acrobat to test files or report on their contents. Finally, you learned how to optimize a file for file size savings.

Exercises

1. Evaluate document collections you are using for various projects. Are there consistent types of document metadata you could add for streamlining file searching?

2. Add a sound file, a movie file, and a Flash file to a document. Experiment with the rendition editing settings, as well as using different actions to initiate the playback of the file.

3. Create a presentation from a group of files on your hard drive. Explore the different transition settings. Is there an optimal time frame to use for the Auto Flip value to advance the slides?

4. Assemble a presentation using Picture Tasks. You need PDF files from JPEG images to use the feature (Chapter 8's project includes one JPEG image named Photo of Field.jpg).

5. Spend some time examining the options and commands available in the Batch Sequence dialog box. Evaluate your own work methods and tasks to see what you could combine into a batch sequence. Then plan and construct the sequence and run it against your files.

6. Construct a Droplet using the method described in the chapter. Some practical options you might like to try are compatibility with a specific Acrobat version, as well as any of a number of image evaluation processes.

7. Using copies of the same file containing images, text, and drawings, experiment with the different optimizing settings in the PDF Optimizer. Compare and contrast output using a variety of settings. What are the differences among the versions? Are the differences significant enough to justify differences in file sizes?

Project

Use the file named **Wow!.pdf** from **ch13_project** folder on the book's Web site as reference for this project.

Task 1: Create the presentation

1. In PowerPoint, make introductory slides, as well as a set of titles you can use for links to files.
2. Convert the file to PDF.
3. Assemble a number of PDF files to use for the presentation.
4. Add links, bookmarks, and buttons to link to your files.

Task 2: Embedding a movie

1. Using your digital video camera, film and edit a movie showing how to work on or fix a piece of equipment described in your presentation.
2. Following the method described in the chapter, embed the movie in a PDF page of your presentation. If you use Acrobat 5 compatibility, the movie is linked to the file; Acrobat 6-compatibility allows the movie to be embedded.
3. Test and save the file.

Task 3: Optimizing the file

Once you have finished the presentation and saved it, take note of the file size. Then use the PDF Optimizer to make the project more compact. Are the differences noticeable visually? In file size?

PDF Mapping – Michael Bufkin

The use of PDF for mapping applications began soon after the first Acrobat products went on sale. Indeed, in his seminal 1991 paper that defined the concept of PDF, "The Camelot Project", John Warnock, then president of Adobe, envisioned a file type that would allow authors of documents to distribute their work in a format that could be accurately displayed and printed on almost any computer or digital printer, including "military maps and complex map collections" [1].

With the advent of the Internet, the advantages of providing users and customers with a single, self-contained file that could be used to display maps which had previously been available only as paper or as complex, proprietary files was obvious. Today, almost all federal, state and local governments are providing mapping data to the public in PDF form, and many commercial map-making enterprises are delivering their map documents in PDF format.

In this Chapter

Working with maps in PDF presents its own challenges and issues. In this chapter you see how to identify these issues and discover means to solve them in discussions and case studies provided by a leader in the PDF mapping field:

- Producing a PDF version of a map from a variety of different types of source material can result in a number of issues, depending both on the source map and the intended use of the PDF map.
- Map PDF files can use the same Acrobat security and navigation features as other types of PDF file.
- Geo-registration is a technique used to embed map projection and coordinate transformation information directly in the PDF.
- Geo-coding is used to associate a latitude and longitude of an address to a bookmark action that zooms to a specific location on a map that has been defined as a PDF coordinate.
- PDF maps may be simple maps, intelligent maps, or map systems using a complex navigation structure for organization and use; best practices are defined for each category.

Issues in Using PDF for Mapping

Publishing a map in PDF can be as simple as selecting the Adobe PDF printer from your Windows print menu if you have Adobe Acrobat installed. However, the resulting PDF may not look very good under close examination. Lines may be very jagged when magnified, the overall size may be that of a letter-sized page, and the colors may not look the way they were intended (Figure 14.1).

There are circumstances when the map's quality may be fine when zoomed out, and be all that is needed to communicate your intent.

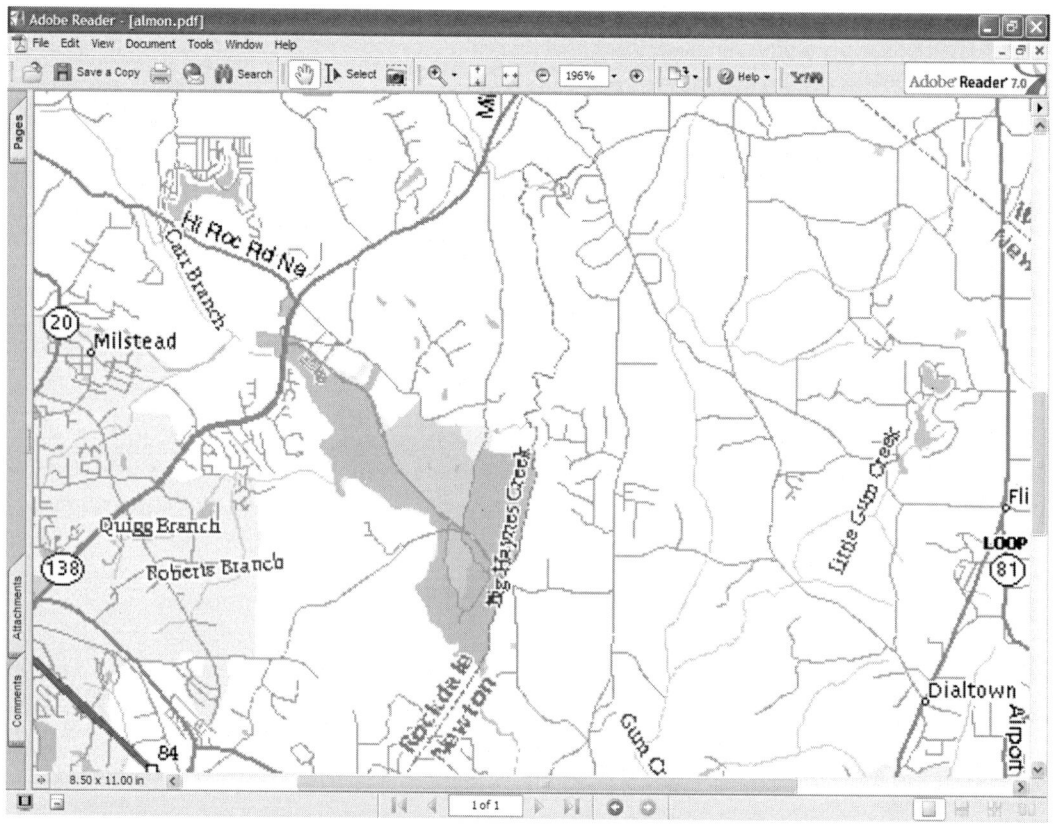

Figure 14.1 Not all PDF maps are created equal

Producing a PDF map that is of the same quality as an equivalent paper print may not be so easy, and taking full advantage of electronic publishing may be even more difficult. The major issues to consider are:

- the source data used to generate the maps
- managing file size
- handling color and printing issues
- working with metadata
- using Acrobat security options
- applying Acrobat navigation features.

Source Data

The first issue to consider is the map's source data. Most maps today are derived from GIS, some of which output directly to PDF. These maps are typically based on a projection and datum which produces at least a base image that is cartographically correct.

Other maps may be made from CAD systems which utilize a set of graphic elements which may or may not have been projected to a Cartesian system before being brought into the CAD environment. Some map source data may simply be scanned images of a paper map, or a satellite image or an aerial photograph georectified to conform to a projection.

Obviously, the quality and type of data will dramatically affect the accuracy and quality of the output map.

File Size

In selecting PDF as a publishing format, one must immediately consider the matter of file size. In general, PDF files are intended to be mobile. They may be made available for users to copy, download, and distribute on CD.

Very large files discourage ease of downloading and distribution. An informal study has shown that 10MB is roughly the maximum file size that people will accept for downloading from the Web. If you want your maps to be used, consider techniques for minimizing file size.

Short of reducing the number of features or the paper size of the map, the two ways to reduce file size are through changes in resolution and compression.

Managing Color and Printing

Another issue of significance in building PDF maps is the handling of color. Different printers may print the same colors differently. In order to have your map display a consistent appearance when printed, it is important to define the colors in the PDF carefully. In particular, if the map is expected to be used for on-demand printing, then color fidelity is extremely important and may require the use of the full capability of Acrobat color management. This may be further complicated by the presence of imagery data, such as satellite data, aerial photographs, or scanned images.

The general issue of printing of maps in PDF form can sometimes be thorny. The use of special fonts can create printing problems, as well as legal issues of font ownership. For maps utilizing PDF layers, it is particularly important to consider the effect of default print settings, as well as the ability of the user to print a map with layer combinations not intended by the author.

PDF Layering

PDF layering is a powerful capability for simplifying the user interface. In particular, layer control by zoom level can significantly improve the display speed of large, highly detailed maps. But it also introduces the possibility of abuse, particularly for documents which may be used in applications which may have legal implications, such as navigation and construction.

Using Metadata

An often overlooked issue of PDF map production is that of metadata. Paper maps may usually include some metadata in the map legend, such as scale, projection, datum, source and date. Electronic maps can carry much more metadata because it takes up no space on the map display. PDF allows an essentially unlimited amount of metadata to be included in the document, and to the extent possible this should be exploited.

Note: Read more about metadata in Chapter 13.

Adding Security

Like metadata, PDF allows the extensive use of security and digital signing techniques to prevent unauthorized use of these documents. They are hidden from document display, but can provide invaluable protections and assurances that the map intent and audience are served without exceeding those parameters.

Using Acrobat Navigation Features

As with many other electronic documents, maps in PDF can take much advantage of standard Acrobat techniques for including bookmarks, hyperlinks and multimedia to generate systems of maps. These can be used to create linked map books which provide navigation between map tiles. Bookmarking can allow users to select from many feature descriptions and automatically zoom to map locations. This can be extended to street addressing, or the location of parcels, equipment, businesses and other features. Such geo-location potential leads to a final set of issues that can provide an enormous extension to the utility of maps in PDF format.

GIS Features for Maps in PDF

Beyond a static representation of a map in a PDF format, it is possible to use advanced cartographic features to make the map a more intelligent document than the standard picture graphic (Figure 14.2).

Today, there are techniques that allow map users to display coordinates, measure distances and directions, query and display feature data, geolocate objects, and connect directly to geospatial databases through PDF forms or ODBC queries.

These advanced features are possible through PDF georegistration, a technique that embeds map projection and coordinate transformation information directly in the PDF. The embedded information then allows any point in PDF coordinates to be transformed directly to the coordinates of the system used to create the map and *vice versa*. Knowing this relationship, any object or record containing a spatial component can be accurately located within the PDF file, and any object in the PDF file can be located in geodetic coordinates.

Georegistration extends the capabilities of a PDF map into areas which were previously only available with complex GIS. If an object in a GIS has associated data, the data can now be associated with the same object in the PDF and can be displayed by the user. Geocoding of addresses is accomplished by associating a latitude and longitude of an address with a Go to a page action assigned to a bookmark. The action zooms to a PDF point which has first been transformed from a lat/long to a PDF coordinate.

The availability of geodetic coordinates can allow outside applications to be used to obtain magnetic declinations, or even point elevations. The combination of georegistered PDF and digital terrain models (DTMs) will eventually allow the draping of the PDF image on the DTM to produce 3D imaging from the map, viewed directly in Adobe Reader.

Best Practices

Given the many different software products used in the production of maps today, it is not possible to provide a detailed guide for each. Maps are being made using products from Adobe, Autodesk, Avenza, Bently Systems, Erdas, ESRI, Intergraph, GE Smallworld, Mapinfo, and others. Some of these products print directly to PDF, most notably AutoCAD, ArcGIS, and Microstation, and this may be the easiest route for those using these products.

With other products the best practice is to output to PostScript or Encapsulated PostScript, and use Adobe Distiller to create the PDF. Even users of products that print directly to PDF may want to use the PostScript option, as Distiller gives much more control over certain output parameters that can impact on PDF fidelity, file size, and performance.

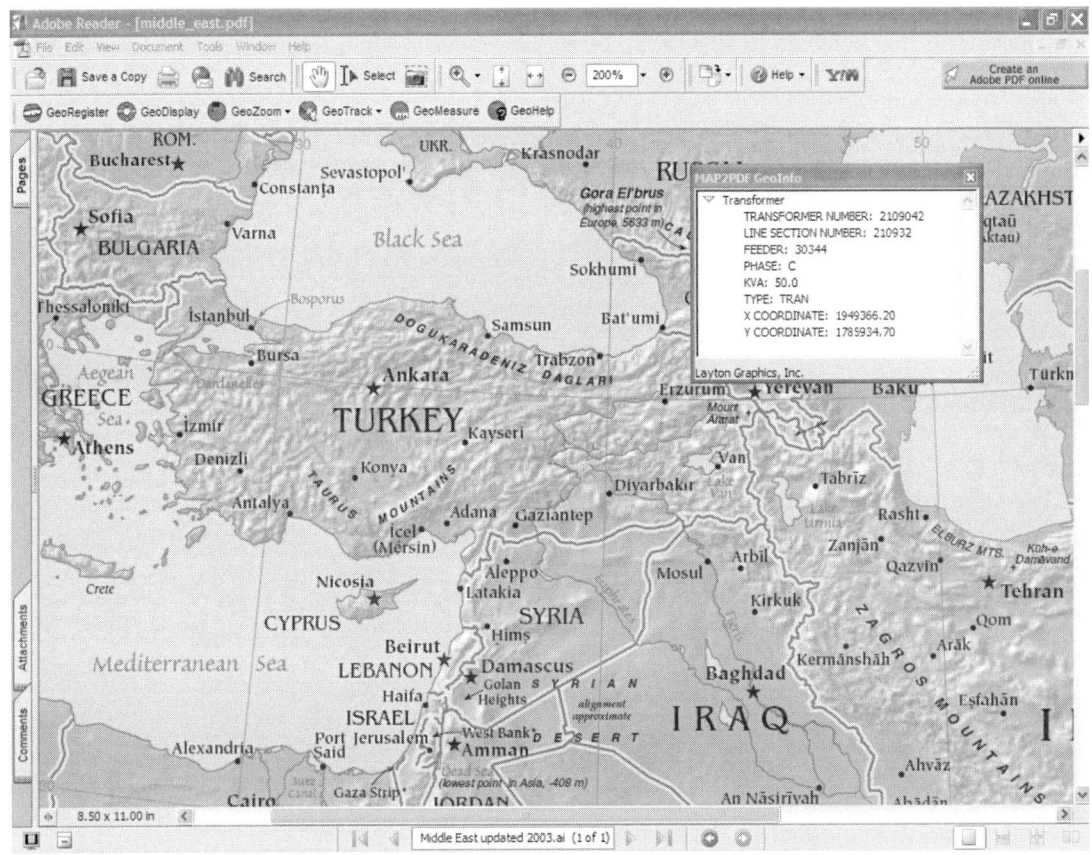

Figure 14.2 A Layton Graphics GeoPDF (GeoPDF © 2005 Layton Graphics. All rights reserved.)

Using Distiller For PDF Map Production

Distiller is a powerful and complex product for converting PostScript and Encapsulated PostScript (EPS) files to PDF. The Distiller Parameters Guide alone is 122 pages, and there are many pages of Distiller references within the Acrobat Help Guide.

Much of the Distiller Guide relates to the use of PDF for high-end printing applications, which may in fact be required for your finished map. Many of the commands and parameters are also useful in creating smaller, faster, better PDF files from PostScript source files.

Distiller allows the user to create and save custom settings, using the dialog box and panels shown in Figure 14.3.

While it is up to the user to determine which settings are best for each use case, here are some starting points:

- **General.** Compatibilityshould be set to Acrobat 6.0 or later. Object Level Compression should be set to Maximum and resolution to no less than 300 dpi. Optimize For Fast Web View should be set if the file is multiple pages and intended for the Web. Fast Web View does not add to speed of a single page file.
- **Images.** For color and gray-scale images, turn Downsample off and set pixels per inch to no less that 300. Compression should be set to JPEG or JPEG2000. There have been reports of display speed problems with JPEG2000 on versions 6 and earlier, so you may want to try both compression methods. In general, a compression quality of Medium or High will give acceptable results.
- **Fonts.** Check both the Embed all fonts and Subset embedded fonts at 100%. Distiller requires that you have permission to use fonts before it will embed them, so be careful in your font choices.
- **Color.** Unless you are knowledgeable in color management, it is suggested that you choose the Leave Color Unchanged option.

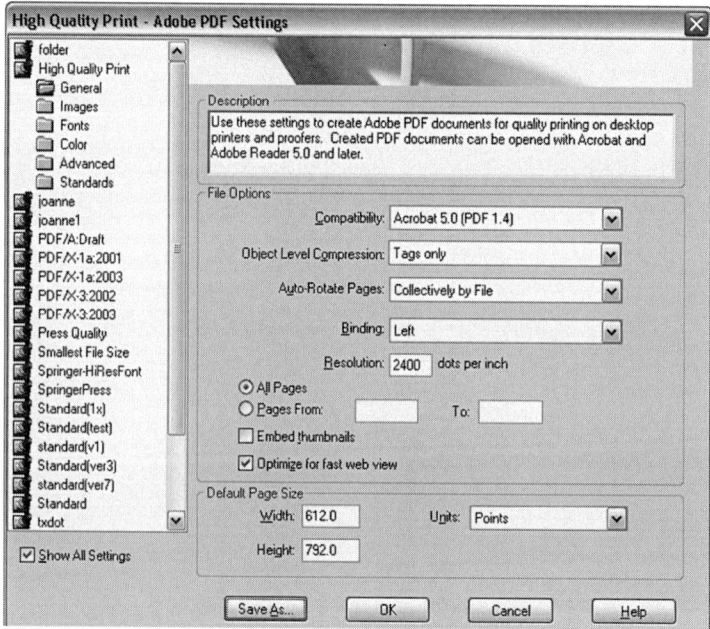

Figure 14.3 Modify Distiller settings as required.

Case Study One: Creating PDF Maps using ESRI ArcMAP

Currently, the most commonly used software for electronic map production is ESRI ArcGIS, using the ArcMAP extension. ArcMAP allows users to set up map frames and systems of maps and then output directly to PDF using the Export map command. Many ESRI users are beginning to use this feature to distribute their map data to their non-technical customers.

The city of Burnaby is a small (40 square mile) but growing municipality west of Vancouver, Canada. Its Engineering Systems group provides mapping data to departmental users and citizens, originally in manually drawn form and more recently through an automated ESRI GIS.

The primary publication medium for this system has been paper prints. The city has now decided to publish the maps in both paper and PDF, with PDF the actual main format, used to print to paper as needed. However, producing quality PDF maps has turned out to be a little more complex than simply selecting the Export to PDF option from the Export menu.

The city chose to use the ESRI DS Mapbook extension, which allows them to create a single multi-page PDF which contains all 286 tiles of the city. Each tile is 500 meters by 800 meters, a size that exactly fills an 11 × 17 page at a scale of 1:2000. While the resulting multi-page PDF is large at about 150 MB, the individual maps are quick to find and display. The single file approach makes it very easy to distribute the entire set, since it only requires the copying of one file.

Goals

Obtaining a compact, quality image that would print reliably on an industry standard 11×17 laser printer was a major goal of the effort. To insure compactness, a dpi of 300 was chosen. It was also decided to use the Universal Condensed font set, which could be closely matched by Adobe's font substitution routines so that the fonts would not have to be embedded. This helped reduce file size, as well as insuring that the PDFs would print consistently, regardless of the printer hardware. Wherever possible, the standard ESRI symbol sets were used to insure consistency and minimal file size.

This product is now being delivered, and the city is looking for ways to improve it. Among techniques being considered are thematic layering, the addition of navigational hyperlinks and the inclusion of feature data attributes. Like many municipalities, the city of Burnaby has just begun to tap the abilities of PDF for distributing their GIS data to the non-technical consumer.

Case Study Two: An Intelligent Map System

North Georgia EMC (NGEMC), an electrical supplier to 95,000 customers in North West Georgia, has implemented a complex map viewing system based on Layton Graphic's GeoPDF technology. NGEMC utilizes a customized AutoCAD system to create and maintain over 1,100 individual maps of their coverage area, using the Georgia West State Plane coordinate system.

Each map contains feature data about the facilities represented on the map in the form of blocks with attribute data. The data contain customer location, transformers, fuses, poles, switches, reclosures, line sections and substations. In addition, there is a separate Customer Information database with information about each current customer.

Converting and Geo-registering the Files

In order to make these data accessible to field and engineering personnel, NGEMC converts each DWG file to PDF, using customized DWG-to-PDF software provided by Layton Graphics. Each DWG is rendered to PDF with only the necessary layers displayed and with a set of custom colors to make the data more readable in Adobe Reader, which uses a white background rather than the standard AutoCAD black background. All title block data, reflecting the map name, the map's position within the overall map grid and revision data are retained (Figure 14.4).

As each PDF is rendered, a post-processing program searches a database for the coordinate points of the corners of the map neatlines and automatically geo-registers the PDF. The processor then searches the associated DWG to determine the state plane coordinates and attributes of each block in the file. The state plane coordinates are transformed to PDF coordinates and the data placed at those points as a custom Adobe annotation. As a user cursors over an annotation, they are alerted to its presence. A single click displays the annotation.

In addition to the attribute data, the grid of maps is displayed in a separate geo-registered PDF index map. Each map sheet name is displayed within the grid. The same database of coordinates is then used to place a hyperlink at each map name location, allowing users to navigate through the entire map set by point and click. This same data are used to create hyperlinks at the edge of each map sheet, which allows users to move between adjacent maps by clicking on the edges.

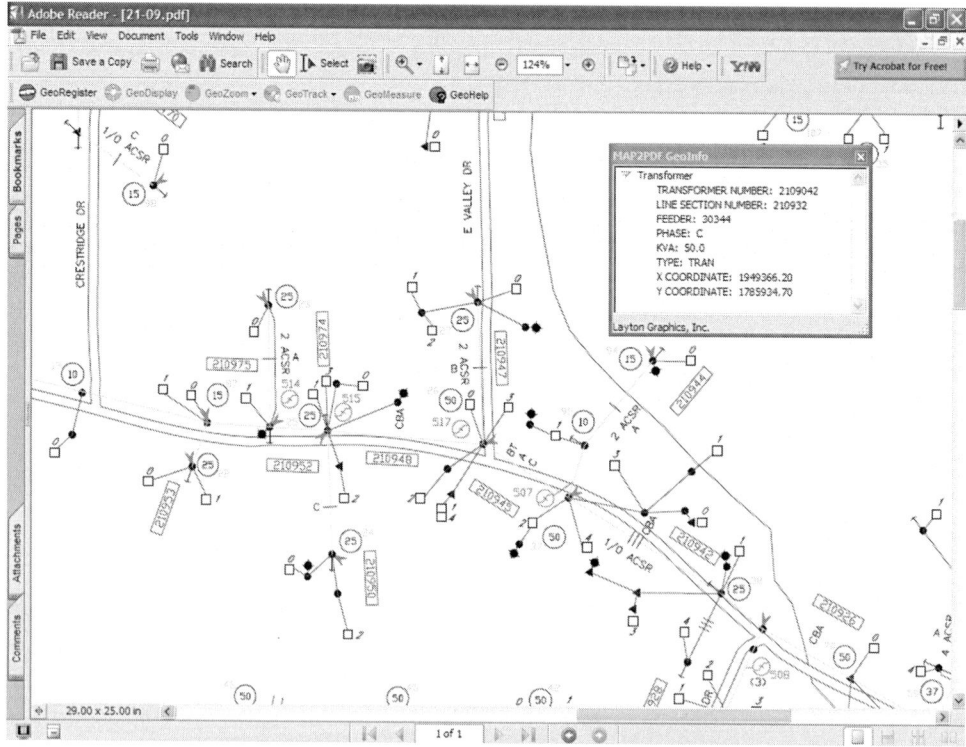

Figure 14.4 GeoPDF with embedded feature data

Hyperlinking Data

Since the index map is geo-registered, address range data, which contains coordinates of street address ranges and street intersections, can be utilized to enable zooming to those locations. The locations are read, and a bookmark for each street intersection and address location is created in the Bookmark pane of the index map.

While rendering the AutoCAD files to PDF and providing them to any computer that has Adobe Reader installed is a great benefit, the additional functionality enabled by the use of Geode is an order of magnitude improvement. What was formerly a simple paper analog has now become a distribution method for the full range of data contained in the NGEMC corporate GIS.

References

[1]   Warnock, John. *The Camelot Project.* 1991. With permission of the author, the PDF document is available at PlanetPDF, http://www.planetpdf.com/mainpage.asp?webpageid=1851.

15

Acrobat for AEC Knowledge Management

One of the major challenges of AEC project management is the tremendous volume of knowledge that has been created in the last 50 years. The time invested in searching for information is a significant productivity drain. PDF and Acrobat offer a way to manage knowledge as never before.

During the 1960s, the United States, under the direction of Dr. W. von Braun, put a man on the moon within a 9-year period using only slide rules, drawing boards, and typewriters. In the 1970s the space program was drastically curtailed, with much of the knowledge scattered among bunkers, garages, government warehouses, microfilm, and within the brains of retirees.

An initiative to return to manned deep-space flight and eventually a manned trip to Mars has made the information valuable again. The New Economy Institute (NEI) is working with the NASA Marshall Space Flight Center under the Congressionally funded Workforce Aging Management Program (WAMP). As a significant percentage of the NASA scientist and engineer populations near retirement, their knowledge needs to be captured, along with other scattered resources.

In this Chapter

In this chapter you see how to use Acrobat's catalog feature to index and create a database of the text in a document collection. You learn how to:

- Collect and store files for ease of indexing
- Build and configure a catalog
- Attach an existing index file to a document
- Search a single document
- Search an index and examine returns using modified and customized search parameters.

The Marshall Star Digital Collection

The Marshall Star [1] is NASA's oldest continuing publication, and is used as a case study throughout this chapter, as well as the chapter's project. The publication, started in fall of 1960, has over 2200 issues and over 16,000 pages. As a result of the WAMP program, all 2200+ issues are now in searchable PDF format. Using Acrobat's search and cataloging features, all pages of the publication can be searched in two seconds or less, for both phrases and ideas.

Organizing the Collection

How do you organize 2,200 documents so an inexperienced user can navigate and search them? One practical solution is to use a navigation page. A table created on a PowerPoint slide is a quick way to generate an attractive page. The navigation page for Marshall is shown in Figure 15.1.

1960	1970	1980	1990	2000	2010
1961	1971	1981	1991	2001	2011
1962	1972	1982	1992	2002	2012
1963	1973	1983	1993	2003	2013
1964	1974	1984	1994	2004	2014
1965	1975	1985	1995	2005	2015
1966	1976	1986	1996	2006	2016
1967	1977	1987	1997	2007	2017
1968	1978	1988	1998	2008	2018
1969	1979	1989	1999	2009	2019

Figure 15.1 Use a table on a single page as an effective navigation page

For scanning, use the Paper Capture (Recognize Text Using OCR in Acrobat 7) process, choosing the Searchable Image Exact format. Each issue was bookmarked by date. To finish the organization for the collection:

1. Place each year in a folder named by year.
2. Link the bookmarked year back to the Navigation page (Figure 15.2).

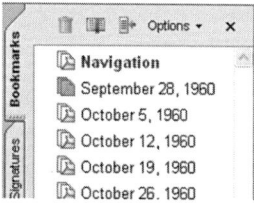

Figure 15.2 Use a navigation bookmark to return the user to the start page

3. To set the document's initial view to show the Bookmarks pane, choose File > Document Properties > Initial View, and choose Open with Bookmark Panel and Page (Figure 15.3).

4. Choose a page layout and magnification option from the dialog box as well. It is sometimes better to choose the open single page and fit width options, as a standard document set to Fit page width is usually unreadable on a computer screen.

5. Create a link from the year on the Navigation Page to the first issue of the year.

Once all documents are prepared, you can make the collection searchable. Acrobat can search a single document or a collection with no preparation. The problem with uncatalogued files is a slower search speed, without the benefit of using the power of metadata in the search. Acrobat can prebuild a fast and very searchable index or Catalog.

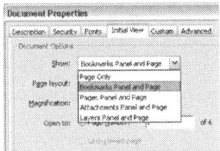

Figure 15.3 Choose how each file opens

Index Planning

A good index can be a great timesaver; a poorly designed index can be very frustrating. Here are some tips for planning your index project:

- Include all directories and files in one location to make it simpler to select content for the index you are building. In addition, keeping all the content together prevents errors.
- Decide whether or not to use *stopwords*, which are common words such as "and" or "the" that can be excluded from the index. You can choose up to 500 case-sensitive words. Although stopwords may produce a slightly smaller index, excluding terms can be confusing when searching for a phrase that includes a stopword. For example, if you specify "and" as a stopword, searching for "Ben and Jerry's" won't produce a result.
- Use custom terms, such as subjects or keywords, when it is useful for your project, which can then be used as search terms. Decide in advance which terms to use as keywords, and which to use as subjects. Do not use the same terms for both. If you search using a subject term and have used it as a keyword in some documents, your search results will be incorrect.
- For your users' convenience, consider building a Readme instruction file, especially if you use custom search terms or stopwords.

Cataloging the Marshall Star

A *catalog* is an electronic index of the words in a series of documents contained in specified folders. Adobe uses the terms *catalog* and *index* almost interchangeably. By using the Catalog function, searches can be greatly customized and search results are returned more quickly than performing an uncataloged search.

Follow these steps to create a catalog:

1. Select Advanced > Catalog from the main program menu to open the Catalog dialog box.

2. Select New Index to start creating the new index; the New Index Definition dialog box opens (Figure 15.4).

3. Name the index using a logical, recognizable name.

4. Select other definition features as required:

- Use an Index Description if numerous indexes are maintained to help users identify the index they want to use.
- Include these directories allows selection of multiple directories to be indexed.
- Exclude these subdirectories allows subfolders to be excluded from the catalog (not shown in the figure.)

5. Click Build to start the process. A dialog box opens for you to define a storage location – Acrobat provides a default location, which is recommended to maintain the index files' integrity.

6. Once the index is complete, the results are listed on the Catalog dialog box (Figure 15.5).

7. Click Close to dismiss the dialog box.

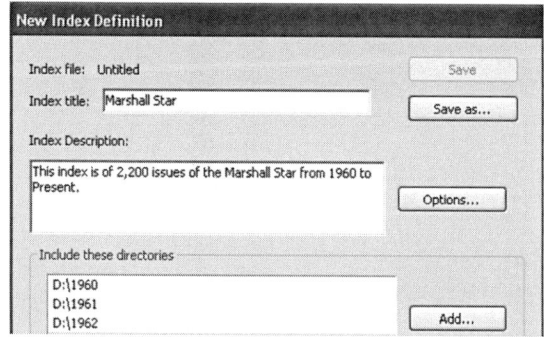

Figure 15.4 Define the contents for the new index

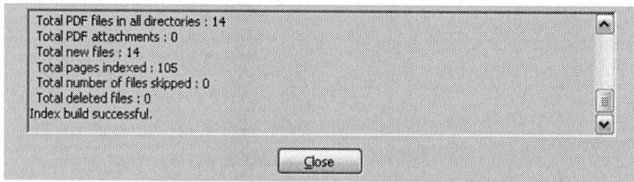

Figure 15.5 The index's characteristics are listed

Associating an Index

The average user does not know how to look for an index; smooth the way by associating an index with the first page a user may encounter. Figure 15.6 shows how to associate an index, in this case, the Marshall Star index.

Follow these steps to associate an index in a file:

1. Choose File > Document Properties and select the Advanced Tab.

2. Click Browse to display the Open dialog box, and locate and select the appropriate index file.

3. Click Open to dismiss the dialog box and attach the index.

4. Click OK to close the Document Properties dialog box. After an index is selected, it is associated with the file after it is saved.

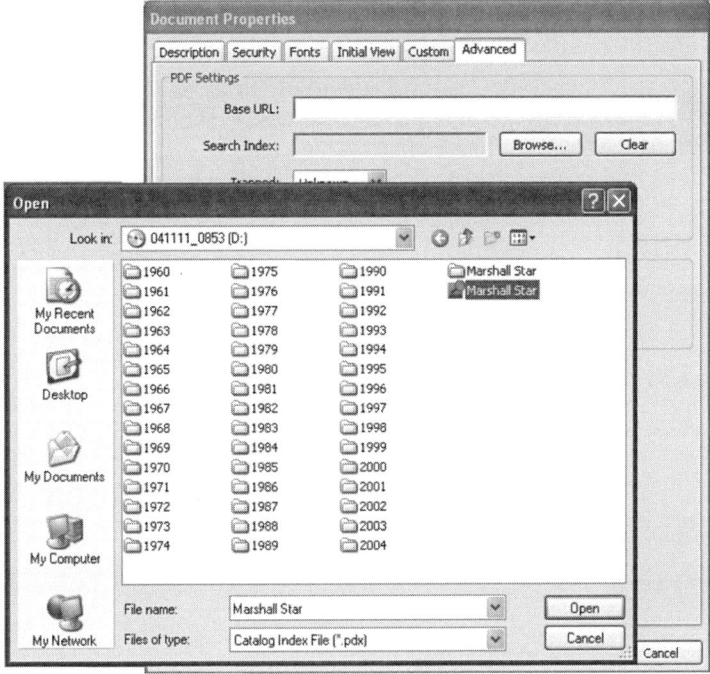

Figure 15.6 Choose an index to associate with a document file

Selecting Multiple Indexes

In both Acrobat versions 6 and 7 you can select multiple indexes. Follow these steps:

1. Choose Search on the Find toolbar, or choose Edit > Search to display the Search PDF window at the right of the program window.

2. Click Use Advanced Search at the bottom of the Search PDF window to display the advanced settings.

3. Click the Look In drop-down arrow and choose Select Index from the list to open the index selection box shown in Figure 15.7.

4. Click Add to locate and select the indexes you want to associate. An index already in use in your system will be listed, click to check the appropriate check box.

5. Click OK to close the dialog box. On the Search PDF window, the Look In field displays the Currently Selected Indexes option.

ARTS Split and Merge Plus

If you scan your documents in groups, such as by year, you can use a third-party plug-in, the ARTS Split and Merge Plus plug-in. The larger files can be split into separate files, each with a complete set of active bookmarks. Delete or store the unsplit file in a different location. Splitting enhances the search ability when using proximity searches.

Uncataloged Searches

Both Acrobat versions 6 and 7 can search large collections of uncataloged documents, whether a single file or across a whole network. Searching a single file is usually done at an acceptable speed unless the file is very large, while searching thousands of files can take a long time.

To conduct an uncataloged search, open the Search PDF window by clicking the Search icon on the File toolbar. Click the All PDF Documents in field'sdrop-down arrow, as shown in Figure 15.8. Select an existing location, or click Browse for Location to find files or folders on your hard drive or network.

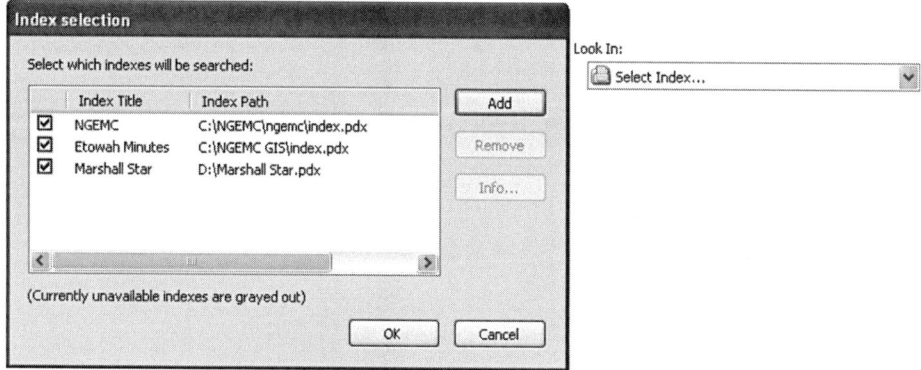

Figure 15.7 Search multiple indexes simultaneously

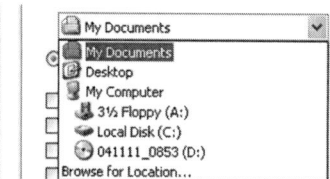

Figure 15.8 Select a folder or storage location to search

Cataloged Searches

Here's a true story: the day the beta of the Marshall Star collection was complete, an engineer called the archives looking for information. He said he knew there were problems with paint on early spacecraft and wanted details.

On paper, the *Marshall Star* fills two large file cabinets with issues, including those with information about the paint problems, but which ones? The first live test was to look for the terms "explorer paint" in the index, as shown in Figure 15.9.

Search Settings

To broaden the search, Match All of the words, Proximity, and Stemming options were also selected. Acrobat's default search proximity is 900 words. In this instance, it was changed to 150 words in the preferences. Choose Edit > Preferences > Search and type a new proximity value, as shown in Figure 15.10. Click OK to close the Preferences dialog box.

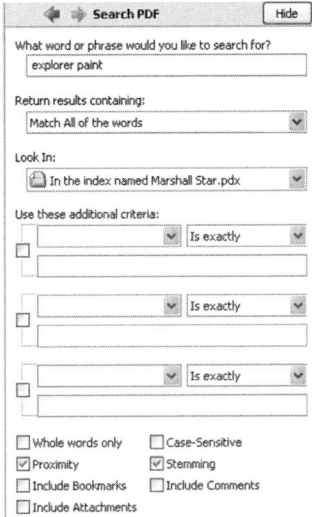

Figure 15.9 The first official query was about early satellites' paint problems

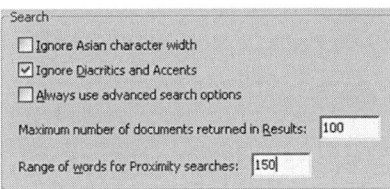

Figure 15.10 Change the proximity value for more accurate search returns

Evaluating Returns

Figure 15.11 shows the search results from the original test with the most relevant results expanded. Acrobat ranks the results by number of hits of the words and how close they are together. In the collection there were 63 documents that had *explorer* and *paint* within 50 words – most of these returns were from later years and in the Employee Classified Ads describing Ford Explorer SUVs with good paint jobs!

Search Outcomes

The document with the highest relevance produced the desired information. Early satellites were painted white, which turned yellow in outer space. Yellow does not reflect heat as well as white, and the internal temperatures in the satellites rose 40 degrees Fahrenheit.

Figure 15.12 shows the new paint scheme on Explorer VII, which was better suited to the space environment. Another search found that the trademark black and white paint job on the Saturn rocket caused metal wrinkling on the second stage from differences of expansion between the black and white painted metals. These searches took seconds instead of days thanks to Acrobat's catalog features.

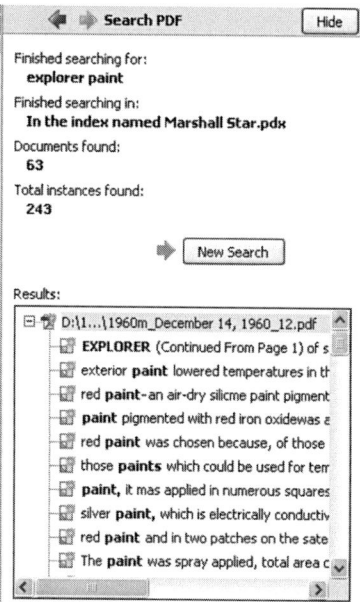

Figure 15.11 Returned search terms are highlighted in the list

Searching and Cataloging Tips

Speed up your searches and examination of results using these tips and hints:
- You cannot search using wildcards such as (?) or (*).
- Tailor the search results. Click the Return results containing field's drop-down arrow and choose from matching all the words, the exact phrase, some of the words, or you can use a Boolean query.
- You can use up to three additional search parameters to customize the search further. Use the drop-down lists below the "Use these additional search criteria" label. Click the left drop-down menu to select the search option, type the term in the field, then click the right drop-down arrow and choose a modifier, such as "Is exactly" or "Is not."
- Sort results using methods other than the file name. Click the Sort by drop-down arrow and choose a method, such as modification date or location.
- Click the Always use Advanced Search option in the Search preferences to automatically open the advanced settings on the Search pane.
- Use the Fast Find preference in the Search preferences to cache the returns from your searches. You can specify the size of the cache, which defaults at 20 MB. Using a cache makes searching faster.

Summary

In this chapter we looked at Acrobat's cataloging process, and its value in the search for information. You saw how to:

- Build and configure an index
- Attach an index to a document
- Conduct and customize searches using both cataloged and uncatalogued documents.

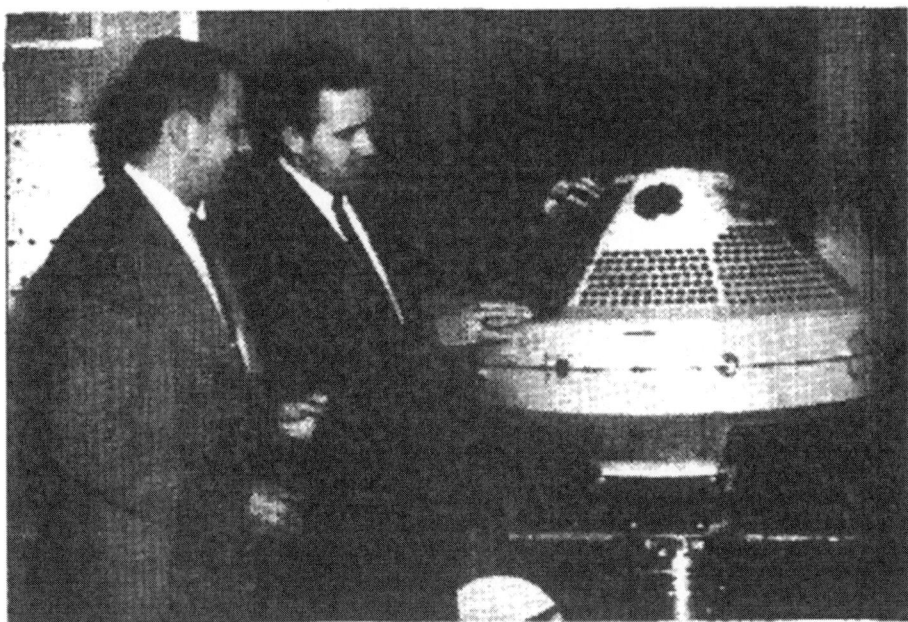

MORE THAN 1,300 one-half inch squares of red iron oxide
paint are helping maintain proper interior temperatures within
the Explorer VIII earth satellite, launched November 3 by the
George C. Marshall Space Flight Center. William C. Snoddy (right)
and Gene A. Zerlaut of the Marshall Center demonstrate the tem-

Figure 15.12 The search took less than two seconds to yield useful results

Exercises

1. Using the methods outlined in the chapter, experiment with a set of documents to create, attach, and search an index.

2. Build a set of stopwords for your test index, and then try searching using excluded terms.

3. Experiment with using keywords and subject terms.

4. Add the terms to the individual documents in your collection, and then rebuild the index. Can you see the value of building a Readme file to organize the specialized search terms?

Project

Using the project files in the **ch15_project** folder on the book's Web site, create and associate an index file with a set of documents. The Catalog folder contains scanned and captured content from the first year of the Marshall Star, NASA's longest-running publication. The file is large and it will take a little time to download over a broadband connection. You will work with the 13 files from 1960, the first year of the publication.

Task 1: Making the index file

In this first task you are going to create a new index file for the document collection.

1. Open **Marshall Star.pdf**, the navigation page for the document collection. When you open the file you may see an error stating that an attached index is missing. As you cannot remove the link to an attached index, the error refers to the original index files. Dismiss the error dialog box, as you will recreate and associate your own index with the page.
2. In Acrobat choose Advanced > Catalog to open the Catalog dialog box.
3. Click New Index to open the New Index Definition dialog box.
4. Type *Marshall Star* in the Index title field; add a description if you like in the Index Description field.
5. Click Add in the Include these Directories section of the dialog box to open a Browse for Folder dialog box. Locate and select the downloaded 1960 folder.
6. Click Build. The Save Index File dialog box opens. Locate the 1960 folder and select it, and click OK to save the index file with the document files.
7. Acrobat processes the files, indexing each word in each file, and displaying the results in the Catalog dialog box.
8. Click Close to dismiss the Catalog dialog box.

Task 2: Associating the index

In this task you attach the new index file to the navigation page PDF for the collection.

1. Open the **Marshal Star.pdf** file in Acrobat.
2. Choose File > Document Properties > Advanced.
3. Click Browse in the PDF Settings section of the dialog box to display an Open dialog box.
4. Locate and select the PDX index file you created in the previous task; click Open to dismiss the dialog box and list the PDX file as the associated index file.
5. Click OK to close the Document Properties dialog box and save the file. Now when the **Marshall Star.pdf** file is opened, the index is automatically attached.

Task 3: Searching the indexed documents

You perform a search in this third task using the features in the Advanced Search Options, which allows your searches to become somewhat fuzzy. In addition to the method described in this task, you can also use additional criteria to conduct very precise searches.

1. Click the Search icon 🔍 on the Find toolbar, or choose Edit > Search to open the Search PDF window.
2. Click Use Advanced Search options to add advanced settings to the Search PDF window.
3. Type *Explorer paint* in the What word or phrase would you like to search for? field. Do not add additional words, such as *and*.
4. Select Match All of the Words in the Return results containing field.
5. The Look In option will state Currently Selected Indexes, as you attached the index in the previous task.
6. In the checkboxes at the bottom of the Search PDF window check Stemming to get paints, painted, painter, un-paintable, and other variations of the search term.

7. Check Proximity to search for returns in documents where the words are within a defined number of words. The default for a proximity search is 900 words, but can be altered in the preferences by choosing Edit > Preferences > Search.
8. Click Search to perform the search. The returns are shown in the Results area of the Search PDF window, and each instance of the terms is highlighted in the list. Your first search return is an article on the Explorer VIII satellite's new red paint job; radiation was changing the original white paint to yellow and raising the internal temperature.

If all the issues of the Marshall Star were searched, you would in later years find that people were selling Ford Explorers with good paint jobs as returns for the search. The Search function will search all 2200 indexed editions as fast as it searches these 13 editions.

It took over an hour to make the 101MB catalog of the Marshall Star, but the search capabilities are extremely powerful and fast. Unlike databases, where we search for exact phrases, we are whole-text searching for ideas. Engineering drawings with searchable text can be cataloged and searched in the same manner.

References

[1] See, for example, the Marshall Web site, http://www.nasa.gov/centers/marshall/home/index.html.

16

Putting It All Together

Throughout this book you have seen how to use Acrobat's features in an AEC project. In this final chapter, we aren't introducing any new information. Rather, we are going to concentrate on building real-life projects.

In our project, the main pages in each navigation file serve as instructional markers, showing you how to proceed and what to include in a specific area. You are invited to use the information in this chapter – and the projects – as a way to develop your own electronic document/knowledge management system.

In this Chapter

In this project chapter you will learn how to use Acrobat for both document and knowledge management, as the processes are the same. There are several elements to the project.

In the project materials available for this chapter, you will find a set of folders matching the navigation structure designed in the project. Each folder includes one PDF file to use as navigation for that folder's contents. The files used in each folder/subfolder are referenced by both folder name and file name.

In addition, you can use the collection of files from the book's main project, the DuPont Soccer Park, also within the chapter's project folder:

- Create the folder structure, and convert files as necessary to PDF and move into the appropriate folders.
- Create a table in a program such as PowerPoint and convert it to PDF to use for a main interface page for navigation.
- Add the links from the main page to a navigation page leading to a subfolder of PDF files.
- Create bookmarks and links from within each area's navigation page.

AEC Document Management

The main interface page for our project is an image of the DuPont Soccer Park with text used for links overlaying the image (Figure 16.1). The original was created in Photoshop. Each of the text labels in the table links to another PDF file used for navigation through a subfolder named according to the section.

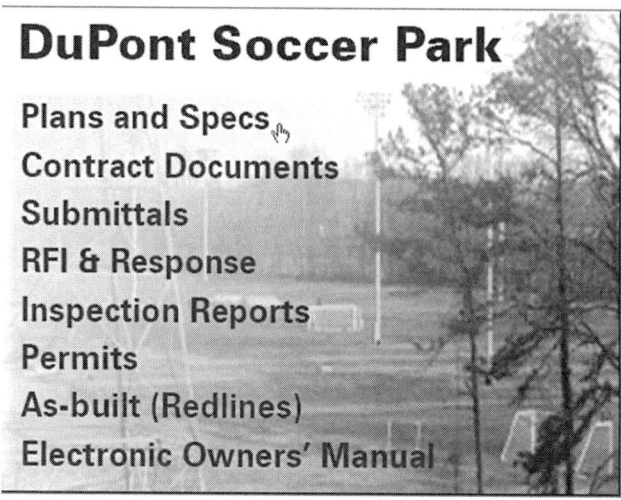

Figure 16.1 The project uses a simple, yet distinctive interface page

Design a folder structure suitable for your project's requirements. In our project, for example, we are creating a document collection for the DuPont Soccer Park. In order to comply with various regulations, collate sources of information, and provide data for the stakeholders, the main interface page of the project leads to eight subfolders.

Before building a project, take the time to develop the folder system you plan to use. Keep in mind that some projects evolve in size and some have varying requirements. However, the general structure of a common AEC project's collection is used in our example. The contents of the subfolders, which also represent a sample navigation path, are shown in Figure 16.2.

Plans and Specs

Just as plans and specifications start a project's development, the Plans and Specs folder is the first linked folder for the document management collection. (Figure 16.3).

Within the main Plans and Specs folder, create a subsystem of folders with varying read – write permissions. For example, assign complete read – write permission for the Civil subfolder to the civil designers, and read permission for the Civil folder to mechanical and electrical designers.

In the PDF file, mirror the folder structure with a bookmark structure (Figure 16.4). That is, design subfolders and corresponding sublevels of bookmarks in the PDF navigation files.

Contracts Documents

Store all project contracts in this folder. In a typical AEC project, this section will include:

- the executed contract and addenda
- change orders
- schedules
- pay requests.

Access to the content in these folders should be restricted. Grant full rights to the accounting and project management team; other project participants won't need to use the information.

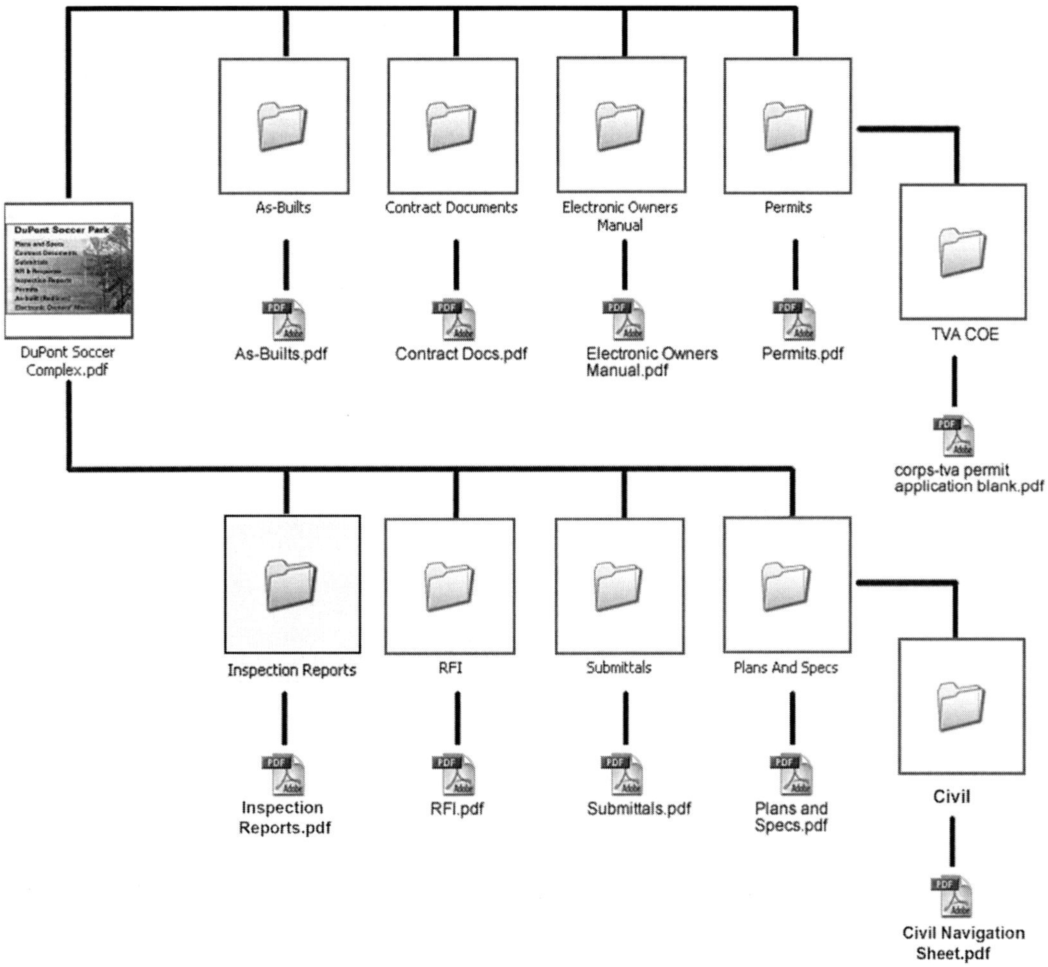

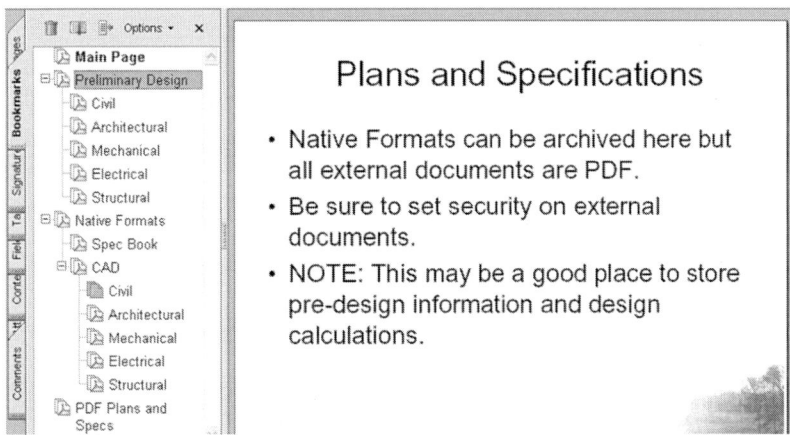

Figure 16.2 The first step is designing the folder structure

Figure 16.3 Plans and Specs are arranged by area and format

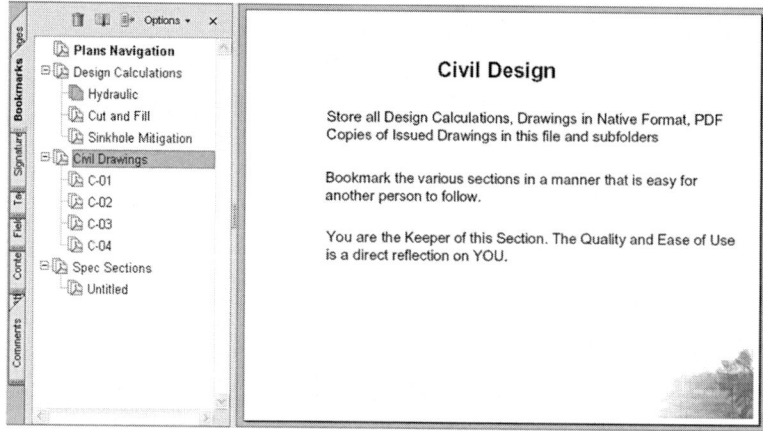

Figure 16.4 Organize content by design area

Submittals

This folder contains the documents submitted by the contractor for approval. When a submittal is received, put it in the folder and bookmark it.

Use the files in this folder for commenting and reviewing. Initiate the review process with the bookmarked file, and the returned comments from the review will be integrated directly into the bookmarked file.

Color Coding

Color coding bookmarks is a handy way to determine the status of a project's submittals from the Submittals PDF page. To make a change in bookmark color, select the bookmark in the Bookmarks pane and then either open the Properties Bar or right-click the bookmark and click Properties to open the Bookmark Properties dialog box. (Figure 16.5).

Our project uses an example of color-coding for submittals:

- a green bookmark means the submittal is accepted
- a red bookmark means the submittal was rejected
- a black bookmark means the submittal is in review
- a purple bookmark means the submittal was conditionally approved.

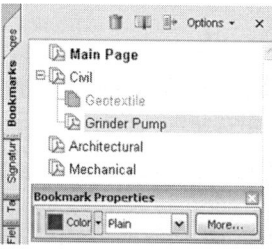

Figure 16.5 Color code bookmarks to illustrate the status of a project element

You can further track the status and progress of submittals using subordinate bookmarks. When the rejected submittals are resubmitted, make the file's bookmark subordinate to the rejected one and start another review with the new file (Figure 16.6).

> **Note:** The first bookmark in each navigation page is a bookmark using colored and bold text linking to the previous navigation page. In this way the user can drill down to a specific drawing or detail, and still make their way back easily to the start of the collection. **NEVER** create a complex structure like this project's without providing the user a way back!

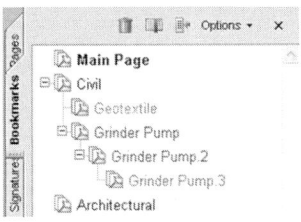

Figure 16.6 Subordinate bookmark levels illustrate submittal status

The (+) to the left of the bookmark identifies both a subordinate bookmark level and a resubmitted submittal. Click the icon to expand the bookmark. When the resubmitted submittal is accepted, change the subordinate bookmark to green; do not replace the initial rejected submittal. Of course, if the resubmission is rejected, proceed with another subordinate bookmark level with the next round of submissions, and maintain the color coding as you progress.

RFI and Response

The Requests for Information and their responses are important legal documents, and are bookmarked to the RFI navigation page from the contents of the the RFI and Response folder. As much of the correspondence may be paper or faxed originally, the documents will require scanning and saving in PDF format. The text of the documents won't need capturing as you are not making any changes to the content.

Archiving Emails

This is also the folder to use for storing emails relating to the project. Before creating the RFI and Response navigation page, define your project's system for maintaining email records. Each person on the project should have subfolders in Outlook's Inbox and Sent box for project-related emails, and create an initial PDF email archive file.

At regular intervals, the emails should be appended to the archive PDF file linked to the RFI and Response subfolder. One of the simplest ways to maintain the content is to use a bookmark linked to each individual's email archive files (Figure 16.7).

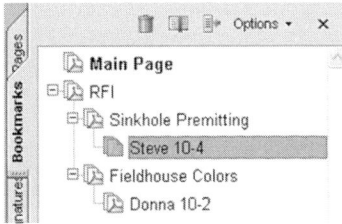

Figure 16.7 Bookmark project participants' email archives

Inspection Reports

All field inspection reports are accessible from the Inspection Reports folder, and bookmarked from the Inspection Reports navigation page by date.

A method that is increasing in popularity is to create the Field Inspection Report as a PDF form that an inspector can fill out using a tablet PC in the field. As well as incorporating the file into the document management collection, deficiencies can also be exported directly from the form to a database or spreadsheet, and a Punch List generated.

Permits

The Permits folder contains scanned copies of all permit applications submitted by the owner, the engineer for the owner, or the contractor, as well as issued permits. For simplicity, add a bookmark naming the permit request, and nest subordinate bookmarks linking to the application, supporting material, and the issued permit.

Like the Submittals navigation page, you can color code bookmarks to differentiate applications from issued permits (Figure 16.8).

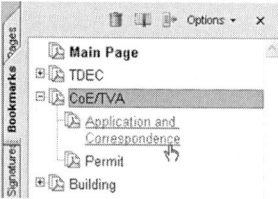

Figure 16.8 Bookmark permits by name

Tip: It is a good idea for the engineer to maintain an as-built set, just in case the contractor omits a change.

As-builts

The As-builts subfolder contains a redlined set of the project drawings linked from the As-builts navigation page. On most projects, the contractor is responsible for the as-builts as a project close-out item.

Electronic Owner's Manual

The final folder contains a single bookmarked and linked PDF that follows the project. The Electronic Owner's Manual (EOM) contains:

- original plans and specs
- owner's manuals for all equipment – each manual should be bookmarked and/or linked
- as-builts
- warranties
- permits (construction and operation)
- construction inspection reports and photographs
- other information that may be needed during the life of the project.

Creating the EOM

Whether the engineer or contractor creates the EOM is a matter of discussion. You can require the contractor to furnish all the Owner's Manuals in bookmarked PDF format, which is usually easier than providing the two or three paper copies now required. Increasingly, owners are willing to pay extra for the EOM, as they can see the benefit and cost-savings over the life of a project.

A project's EOM will be a rather large file, and can easily reach 3,000 pages. For testing purposes we created a 10,279-page (1.03GB) file. The file worked well on a computer with a 1.7 GHz processor using 512MB of RAM. Although creating a document containing many files and pages seems like a daunting task, using the Create PDF from Multiple Files command in Acrobat 6 and 7 makes easy work of assembling the file.

Building a Knowledgebase

A *knowledgebase* is an organized collection of what is known on a subject. The document management project described earlier for the AEC project is a knowledgebase for that project. All engineering companies need a knowledgebase of regulations, zoning maps, flood maps, permit application forms, design standards, and so on. Making a collection seems like a lot of work – and in fact it IS a lot of work – but its advantages outweigh the time and effort involved in the construction. The finished document can be easily distributed as needed, saving duplication and delivery time and costs.

A Case in Point

In addition to working for the NEI, Tom also works with the Southeast Tennessee Development District (SETDD). The SETDD, all 10 counties in the district, and the cities in the counties, need Federal Emergency Management Agency flood maps. Most entities had paper copies, but generating and delivering paper maps incurred significant costs.

Tom converted TIFF images of all the maps to PDF. The task took about eight hours in total to download, link, and bookmark all 300+ maps. However, now the file is complete, all the government offices can have a set of maps electronically in minutes. Instead of copying and faxing, requests from the public for sections of the flood map can be emailed in seconds.

Most engineering companies, municipalities, industries, school districts, and other similar organizations often have a dungeon-like room used to store old drawings. The drawings must be maintained, and on the occasion when a particular drawing is required, considerable time is invested in searching for the drawing.

Numerous clients are converting the archives to scanned PDF drawings, thereby freeing up the storage space, not to mention removing the fire hazard. The New Economy Institute and the Tennessee Valley Authority run a joint program where high-school seniors are paid to convert scanned drawings into CAD files. The students use large-format scanners to scan old plans to PDF and store the PDF files in the corresponding government's knowledgebase.

Challenges

A knowledgebase is infinitely expandable. As more content needs to be added, you simply modify the navigation pages and add more folders. In your organization, give the most appropriate person or persons the drive space and the mission and turn them loose. Use the design as a competition, either formal or informal, and award the best developed areas in the knowledgebase.

The Next Step

Acrobat provides the AEC professional with straight-forward tools to increase productivity dramatically. Since PDF is very close to a paper-based workflow, it is simple to make the transition to the electronic model.

Appendix A

System Requirements

On Windows, Acrobat 7 Professional can operate using Windows 2000 with Service Pack 2, Windows XP Professional, XP Home or Tablet PC Edition. In addition, your system requires:

- Intel Pentium-class processor
- 128 MB of RAM (256 MB recommended)
- 385 MB of available hard-disk space; optional installation files cache (recommended) requires an additional 385 MB of available hard-disk space
- Microsoft Internet Explorer 5.5
- 800 × 600 screen resolution monitor
- CD-ROM drive
- Internet connection or phone call required for product activation

To run Acrobat 7 Professional on Mac OS requires:

- PowerPC G3, G4 or G5 processor
- Mac OS X v.10.2.8, 10.3
- 128 MB of RAM (256 MB recommended)
- 450 MB of available hard-disk space
- 800 × 600 screen resolution monitor
- CD-ROM drive

Appendix B

Acrobat 7 and Adobe Reader 7 Accessibility

Acrobat 7 and Adobe Reader 7 include accessibility features for creating, evaluating, and reading documents that comply with mandates, such as Section 508 of the US Rehabilitation Act.

This appendix provides an introductory overview to the functions included with Acrobat 7 and Adobe Reader 7; please consult the appropriate program's Help files for full information.

Accessibility Features

Acrobat 7 Professional offers a number of tools and processes that help you create and optimize PDF files for use as accessible documents. The features include:

- **Add Tags.** Use the Add Tags feature to evaluate the structure of a document and provide a set of tags to use for control and display of the content.
- **Accessibility Checker.** This feature evaluates a document and adds tags to an untagged document. The results of a check are displayed in the Accessibility Checker log displayed in the How To window. The report includes listings of missing tags, such as <alt> tags for images, as well as other possible optimizations.
- **TouchUp Reading Order.** Use the TouchUp Reading Order tool and dialog boxes to modify the tag structure on a page and define a logical reading order.

Accessibility Setup Assistant

Both Acrobat 7 and Adobe Reader 7 include a five-pane wizard used to set up the program for use with a screenreader or magnifier. Each screen of the wizard allows you to make decisions about some aspect of the accessibility setup without having to change preference settings in a number of dialog boxes.

In Acrobat 7, choose Advanced > Accessibility > Setup Assistant; in Adobe Reader 7, choose Help > Accessibility Setup Assistant to open the wizard.

Note: The wizard is the same in both Acrobat and Adobe Reader, with the exception of references to the program's name, which vary according to which program you are working in.

Make configuration choices on the successive panes of the wizard as follows:

- **Screen 1.** Choose the device you are working with, either a screen reader, screen magnifier, or both. From this pane you can also click Use recommended settings and skip setup check box to use Acrobat's preconfigured settings.
- **Screen 2.** Choose a high-contrast color scheme, text smoothing option, and a default zoom for document viewing. Check the Always use the keyboard selection cursor check box, depending on your device.
- **Screen 3.** Choose options for tagging. The default allows the program to interpret the reading order, but you can also specify another option to override or confirm tagging.
- **Screen 4.** Choose options for viewing large documents, defined as 50 pages by default. Choose from the visible page, entire document, or let the program decide.
- **Screen 5.** Specify opening and saving document settings, including the auto-save.

Index